Lecture Notes in Computational Science and Engineering

129

Editors:

More information about this series at http://www.springer.com/series/3527

Michael Griebel • Marc Alexander Schweitzer
Editors

Meshfree Methods for Partial Differential Equations IX

Springer

Editors
Michael Griebel
Institut für Numerische Simulation
Universität Bonn
Bonn, Germany

Fraunhofer-Institut für Algorithmen
und Wissenschaftliches
Rechnen SCAI,
Sankt Augustin, Germany

Marc Alexander Schweitzer
Institut für Numerische Simulation
Universität Bonn
Bonn, Germany

Fraunhofer-Institut für Algorithmen
und Wissenschaftliches
Rechnen SCAI,
Sankt Augustin, Germany

ISSN 1439-7358 ISSN 2197-7100 (electronic)
Lecture Notes in Computational Science and Engineering
ISBN 978-3-030-15121-8 ISBN 978-3-030-15119-5 (eBook)
https://doi.org/10.1007/978-3-030-15119-5

Mathematics Subject Classification (2010): 65N30, 65N75, 65M60, 65M75, 65Y99

Cover illustration: Image reprinted with kind permission from Albert Ziegenhagel and Matthias Birner.

This Springer imprint is published by the registered company Springer Nature Switzerland AG.
The registered company address is: Gewerbestrasse 11, 6330 Cham, Switzerland

Preface

The Ninth International Workshop on *Meshfree Methods for Partial Differential Equations* was held from September 18 to September 20, 2017, in Bonn, Germany. Meshfree methods have a diverse and rich mathematical background and their flexibility renders them particularly interesting for challenging applications in which classical mesh-based approximation techniques struggle or even fail. This workshop series was established in 2001 to bring together European, American, and Asian researchers working in this exciting field of interdisciplinary research on a regular basis.

To this end, Ivo Babuška, Jiun-Shyan Chen, Michael Griebel, Wing Kam Liu, Marc Alexander Schweitzer, C. T. Wu, and Harry Yserentant invited scientists from all over the world to Bonn to strengthen the mathematical understanding and analysis of meshfree discretizations and to promote the exchange of ideas on their implementation and application.

The workshop was again hosted by the Institut für Numerische Simulation at the Rheinische Friedrich-Wilhelms-Universität Bonn with the financial support of the Sonderforschungsbereich 1060 *The Mathematics of Emergent Effects* and the Hausdorff Center for Mathematics.

Bonn, Germany — Michael Griebel
Bonn, Germany — Marc Alexander Schweitzer
December 2018

Contents

Preconditioned Conjugate Gradient Solvers for the Generalized Finite Element Method

Travis B. Fillmore, Varun Gupta, and Carlos Armando Duarte

Abstract This paper focuses on preconditioners for the conjugate gradient method and their applications to the Generalized FEM with global-local enrichments (GFEM^{gl}) and the Stable GFEM^{gl}. The preconditioners take advantage of the hierarchical struture of the matrices in these methods and the fact that most of the matrix does not change when simulating for example, the evolution of interfaces and fractures. The performance of the conjugate gradient method with the proposed preconditioner is investigated. A 3-D fracture problem is adopted for the numerical experiments.

1 Introduction

The Generalized or Extended FEM (GFEM/XFEM) [3, 4, 6, 12, 26, 28, 31] has successfully been applied to problems involving moving interfaces, crack propagation, material discontinuities, and many others. These applications rely on a-priori knowledge of the solution in order to define enrichment functions. Several assumptions are usually required for the derivation of these enrichments. As a result, refinement of the FEM mesh is usually required for acceptable accuracy. One strategy to address this issue is to define the enrichments numerically as the solution of auxiliary boundary value problems [10]. This leads to the so-called Generalized FEM with global-local enrichments (GFEM^{gl}). Another limitation of the GFEM is the ill-conditioning of the resulting system of equations which may lead to severe round-off errors of direct solvers or to the lack convergence of iterative solvers.

T. B. Fillmore · C. A. Duarte (✉)
Department of Civil and Environmental Engineering, University of Illinois at Urbana-Champaign, Urbana, IL, USA
e-mail: travis.b.fillmore@usace.army.mil; caduarte@illinois.edu

V. Gupta
Pacific Northwest National Laboratory, Richland, WA, USA
e-mail: varun.gupta@pnnl.gov

M. Griebel, M. A. Schweitzer (eds.), *Meshfree Methods for Partial Differential Equations IX*, Lecture Notes in Computational Science and Engineering 129,
https://doi.org/10.1007/978-3-030-15119-5_1

Preconditioning schemes for the GFEM can be found in the works of Kim et al. [23] which proposes a Block-Jacobi preconditioner for the conjugate gradient method, Waisman et al. [7, 36], Menk and Bordas [27], Béchet et al. [5], and several others.

The Stable GFEM (SGFEM), initially proposed in [1, 2] and extended to 2- and 3-D fracture mechanics in [17, 18], provides a robust and yet simple solution to the problem of ill-conditioning of the GFEM/XFEM. It is shown in [2] that the SGFEM yields matrices with a condition number which is orders of magnitude lower than in the GFEM/XFEM.

Block Gauss-Seidel iterative solution algorithms for the SGFEM are proposed in Kergrene et al. [20] and [15]. This paper proposes preconditioners for the SGFEM and, in particular to the SGFEM^{gl} –an application of SGFEM ideas to the GFEM^{gl} which was first proposed in [15]. The key idea of the preconditioners is to explore the hierarchical structure of the system of equations in Generalized FEMs like the SGFEM^{gl}. For example, when simulating the propagation of fractures in a domain using the SGFEM^{gl}, only the enrichments change between propagation steps. The entries of the matrix associated with the FEM space—a sub-space of the SGFEM^{gl} space—remains constant throughout the entire simulation regardless of the complexity of the fracture. This has been demonstrated in [32]. Two preconditioners for the conjugate gradient method are investigated: The Block Jacobi (BJ-PCG) and the Block Gauss-Seidel (BGS-PCG). They are defined in Sect. 4 and their performance investigated in Sect. 5. The numerical experiments involve the solution of a 3-D fracture problem using the SGFEM^{gl}. This method is briefly reviewed in Sect. 3. The model problem adopted in this paper—the linear elastic fracture mechanics problem—is summarized in Sect. 2. The main conclusions of this work are presented in Sect. 6.

2 Model Problem

The iterative solvers investigated in this paper are not restricted to a particular problem. However, we focus on linear elastic fracture mechanics problems in 2- and 3-D. Consider a cracked domain $\bar{\Omega} = \Omega \cup \partial\Omega$ in $\mathbb{R}^d$, $d = 2$ or 3, as illustrated in Fig. 1. The boundary is decomposed as $\partial\Omega = \partial\Omega^u \cup \partial\Omega^\sigma$ with $\partial\Omega^u \cap \partial\Omega^\sigma = \emptyset$. The crack surface $S \subset \partial\Omega^\sigma$ is assumed to be traction-free. We consider the linear elasticity problem on this domain. The equilibrium equations are given by

$$\nabla \cdot \boldsymbol{\sigma} = \mathbf{0} \qquad \text{in } \Omega, \tag{1}$$

where $\boldsymbol{\sigma}$ is the Cauchy stress tensor. The following boundary conditions are prescribed on $\partial\Omega$

$$\boldsymbol{u} = \bar{\boldsymbol{u}} \text{ on } \partial\Omega^u \qquad \boldsymbol{\sigma} \cdot \boldsymbol{n} = \bar{\boldsymbol{t}} \text{ on } \partial\Omega^\sigma, \tag{2}$$

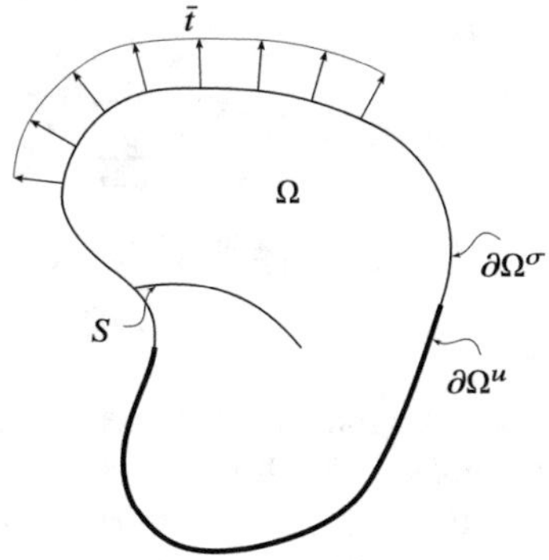

Fig. 1 Fractured domain $\bar{\Omega}$ in $\mathbb{R}^2$ or $\mathbb{R}^3$

where $\boldsymbol{n}$ is the outward unit normal vector to $\partial\Omega^\sigma$ and $\bar{\boldsymbol{t}}$ and $\bar{\boldsymbol{u}}$ are prescribed tractions and displacements, respectively. Without loss of generality, we assume hereafter that $\bar{\boldsymbol{u}} = \boldsymbol{0}$. The constitutive relations are given by the generalized Hooke's law,

$$\boldsymbol{\sigma} = \boldsymbol{C} : \boldsymbol{\varepsilon}, \tag{3}$$

where $\boldsymbol{C}$ is Hooke's tensor. The kinematic relations are given by

$$\boldsymbol{\varepsilon} = \nabla_s \boldsymbol{u} \qquad \text{in } \Omega, \tag{4}$$

where $\boldsymbol{\varepsilon}$ is the linear strain tensor and ∇_s is the symmetric part of the gradient operator. We seek to find a GFEM approximation to the solution $\boldsymbol{u}$ of the problem defined by Eqs. (1)–(4).

3 GFEM and GFEM^{gl} Approximations

A brief review of generalized FEM approximations is given in this section. Further details can be found in, for example, [3, 11, 26, 31, 34].

The GFEM test and trial space $\mathbb{S}^{\text{GFEM}}$ is obtained by hierarchically enriching a low-order standard finite element approximation space $\mathbb{S}^{\text{FEM}}$, with special functions related to the given problem and belonging to the enrichment space $\mathbb{S}^{\text{ENR}}$. Consider a finite element mesh covering the domain of interest $\bar{\Omega}$. Let $N_\alpha(\boldsymbol{x})$, $\alpha \in I_h = \{1, \cdots, \text{nnod}\}$, be the standard linear finite element shape function associated with node $\boldsymbol{x}_\alpha$ and with support ω_α. The *patch or cloud* ω_α is given by the union of the finite elements sharing node $\boldsymbol{x}_\alpha$. The test/trial space of the GFEM is given by

$$\mathbb{S}^{\text{GFEM}} = \mathbb{S}^{\text{FEM}} + \mathbb{S}^{\text{ENR}}, \tag{5}$$

where

$$\mathbb{S}^{\text{FEM}} = \sum_{\alpha \in I_h} \underline{\hat{\boldsymbol{u}}}_\alpha N_\alpha, \quad \underline{\hat{\boldsymbol{u}}}_\alpha \in \mathbb{R}^d,\ d = 2, 3,$$

$$\mathbb{S}^{\text{ENR}} = \sum_{\alpha \in I_h^e} N_\alpha \chi_\alpha, \text{ and } \quad \chi_\alpha(\omega_\alpha) = \text{span}\{E_{\alpha i}\}_{i=1}^{m_\alpha}. \tag{6}$$

The basis function $E_{\alpha i}$ is called an *enrichment function*, $\alpha \in I_h^e \subset I_h$ is the index of the node with this enrichment, and $i = \{1, \cdots, m_\alpha\}$ is the index of the enrichment function at the node with m_α being the total number of enrichments associated with node $\boldsymbol{x}_\alpha$. The functions $E_{\alpha i} \in \chi_\alpha(\omega_\alpha)$ are chosen such that they approximate the unknown solution $\boldsymbol{u}$ of the problem locally in ω_α. Examples of enrichment functions are polynomials, the Heaviside function, crack tip singular functions, and numerically generated functions (cf. Sect. 3.1). The spaces $\chi_\alpha(\omega_\alpha)$ are called *patch* approximation spaces, and $\mathbb{S}^{\text{ENR}}$ is referred to as the *global enrichment space* of the GFEM. The functions in $\mathbb{S}^{\text{ENR}}$

$$\phi_{\alpha i}(\boldsymbol{x}) = N_\alpha(\boldsymbol{x}) E_{\alpha i}(\boldsymbol{x}), \qquad \alpha \in I_h^e,\ i = 1, \ldots, m_\alpha, \tag{7}$$

are denoted GFEM shape functions. They are built from the product of Finite Element shape functions, $N_\alpha(\boldsymbol{x})$, $\alpha \in I_h^e$, and enrichment functions, $E_{\alpha i}$, $i = 1, \ldots, m_\alpha$. There are m_α GFEM shape functions at a node $\boldsymbol{x}_\alpha$, $\alpha \in I_h^e$, of a finite element mesh. These nodes also have a standard FE shape function $N_\alpha \in \mathbb{S}^{\text{FEM}}$. Nodes not in the set I_h^e have only one function—the FE shape function N_α.

Based on Eqs. (5)–(7), the GFEM approximation $\boldsymbol{u}^{\text{GFEM}}$ of a vector field $\boldsymbol{u}$ (e.g., displacements) can be written as

$$\begin{aligned}\boldsymbol{u}^{\text{GFEM}}(\boldsymbol{x}) &= \boldsymbol{u}^{\text{FEM}}(\boldsymbol{x}) + \boldsymbol{u}^{\text{ENR}}(\boldsymbol{x}) \\ &= \underbrace{\sum_{\alpha \in I_h} \underline{\hat{\boldsymbol{u}}}_\alpha N_\alpha(\boldsymbol{x})}_{\text{Standard FEM approx.}} + \underbrace{\sum_{\alpha \in I_h^e} N_\alpha(\boldsymbol{x}) \sum_{i=1}^{m_\alpha} \underline{\tilde{\boldsymbol{u}}}_{\alpha i} E_{\alpha i}(\boldsymbol{x})}_{\text{GFEM enriched approx.}}, \quad \underline{\hat{\boldsymbol{u}}}_\alpha, \underline{\tilde{\boldsymbol{u}}}_{\alpha i} \in \mathbb{R}^d,\ d = 2, 3.\end{aligned} \tag{8}$$

The above equation clearly shows that a GFEM approximation is obtained by hierarchically enriching a standard finite element approximation. As a consequence, any GFEM stiffness matrix is given by a FEM matrix augmented with entries associated with GFEM enrichments. This property of GFEM matrices is used in the proposed preconditioners for the GFEM.

3.1 The Generalized FEM with Global-Local Enrichments

Available enrichment functions for linear elastic fracture problems [6, 12, 13, 28, 30] are based on the expansion of the elasticity solution in the neighborhood of a straight crack front in an infinite domain. They also assume a planar fracture surface. These assumptions are not valid in most practical fracture mechanics problems, in particular for the case of 3-D problems. As a result, refinement of the FEM mesh is required for acceptable accuracy. Alternatively, the enrichments can be defined numerically as the solution of auxiliary boundary value problems [10, 14]. This so-called Generalized FEM with Global-Local Enrichments (GFEM^{gl}), combines the GFEM and the global-local FEM [9, 29]. This allows the GFEM to use coarse meshes while delivering accurate solutions. The GFEM^{gl} has been formulated and applied to various classes of problems. In Sect. 5, the method is used to discretize a 3-D linear elastic fracture problem. The resulting discrete system of equations is solved using the iterative solvers described in Sect. 4. Further details on GFEM^{gl} in the context of linear elastic fracture mechanics, can be found in [10, 21, 22].

3.2 Stable GFEM and Stable GFEM^{gl}

It can be shown that the growth of the condition number for the GFEM is $O(h^{-4})$ with mesh refinement [2]. In contrast, the condition number of the *standard* FEM stiffness matrix for a 3-D elasticity problem subjected to Neumann boundary conditions is $O(h^{-2})$ [8]. It is noted that if point constraints are adopted in 3-D to eliminate the rigid body motions, the condition number of the FEM matrix with these point constraints is $O(h^{-3})$ [8]. Condition number in this paper is taken to mean the condition number computed using the non-zero eigenvalues of a matrix scaled such that its diagonal entries are 1 or $O(1)$. This is also known as the scaled condition number.

The Stable GFEM (SGFEM) [2] was proposed to address this ill-conditioning issue of the GFEM. In the SGFEM, the enrichment functions are locally modified to construct the patch approximation spaces $\widetilde{\chi}_\alpha$, $\alpha \in I_h^e$. The modified SGFEM enrichment functions $\tilde{E}_{\alpha i}(\boldsymbol{x}) \in \widetilde{\chi}_\alpha(\omega_\alpha)$ are given by

$$\tilde{E}_{\alpha i}(\boldsymbol{x}) = E_{\alpha i}(\boldsymbol{x}) - \mathrm{I}_{\omega_\alpha}(E_{\alpha i})(\boldsymbol{x}) \text{ and } \widetilde{\chi}_\alpha = \operatorname{span}\{\tilde{E}_{\alpha i}\}_{i=1}^{m_\alpha} \tag{9}$$

where $\mathrm{I}_{\omega_\alpha}(E_{\alpha i})$ is the piecewise linear finite element interpolant of the enrichment function $E_{\alpha i}$ on the patch ω_α. The global enrichment space associated with $\widetilde{\chi}_\alpha$ is denoted by $\widetilde{\mathbb{S}}^{\mathrm{ENR}}$. Therefore, the SGFEM trial space $\mathbb{S}^{\mathrm{SGFEM}}$ is given by

$$\mathbb{S}^{\mathrm{SGFEM}} = \mathbb{S}^{\mathrm{FEM}} + \widetilde{\mathbb{S}}^{\mathrm{ENR}}. \tag{10}$$

The SGFEM shape functions $\tilde{\phi}_{\alpha i}(\boldsymbol{x})$ belonging to $\widetilde{\mathbb{S}}^{\mathrm{ENR}}$ are constructed using the same framework as GFEM and are given by

$$\tilde{\phi}_{\alpha i}(\boldsymbol{x}) = N_{\alpha}(\boldsymbol{x})\tilde{E}_{\alpha i}(\boldsymbol{x}). \tag{11}$$

The above procedure can also be applied to the GFEM^{gl} [15, 25]. The resulting methodology is denoted as the SGFEM^{gl}. Further details about the SGFEM are given in [1, 2, 17]. The numerical implementation of the SGFEM is described in Section 4 of [17].

4 Iterative Solvers

The iterative solvers studied in this paper are the Block Gauss-Seidel (BGS), the Conjugate Gradient (CG), Block Jacobi Preconditioned CG (BJ-PCG), and Block Gauss-Seidel PCG (BGS-PCG). All of the "Block" iterative solvers take advantage of the hierarchical nature of the GFEM/SGFEM approximation spaces (5) and (10). This property leads to the following structure for the global stiffness matrix $\boldsymbol{K}$, displacement vector $\boldsymbol{d}$, and load vector $\boldsymbol{f}$ associated with a GFEM/SGFEM discretization of the problem described in Sect. 2:

$$\boldsymbol{K}\boldsymbol{d} = \begin{bmatrix} \boldsymbol{K}^{0} & \boldsymbol{K}^{0,\mathrm{gl}} \\ \boldsymbol{K}^{\mathrm{gl},0} & \boldsymbol{K}^{\mathrm{gl}} \end{bmatrix} \begin{bmatrix} \boldsymbol{d}^{0} \\ \boldsymbol{d}^{\mathrm{gl}} \end{bmatrix} = \begin{bmatrix} \boldsymbol{f}^{0} \\ \boldsymbol{f}^{\mathrm{gl}} \end{bmatrix} = \boldsymbol{f}, \tag{12}$$

where $\boldsymbol{K}^0$ is associated with the FEM space $\mathbb{S}^{\mathrm{FEM}}$, $\boldsymbol{K}^{\mathrm{gl}}$ is associated with enrichment space $\mathbb{S}^{\mathrm{ENR}}$ or $\widetilde{\mathbb{S}}^{\mathrm{ENR}}$, and $\boldsymbol{K}^{\mathrm{gl},0} = (\boldsymbol{K}^{0,\mathrm{gl}})^{\mathrm{T}}$ represents the coupling between the FEM and enrichment spaces.

Remark 1 The notation $\boldsymbol{K}^{0,gl}$, $\boldsymbol{K}^{gl}$, $\boldsymbol{d}^{\mathrm{gl}}$, and $\boldsymbol{f}^{gl}$ is adopted since the enrichments used in this paper are computed through a global-local analysis as described in Sect. 3.1.

Remark 2 Matrix $\boldsymbol{K}^0$ does not change in a crack propagation simulation. Thus, it can be factorized once and re-used to define an efficient pre-conditioner for the GFEM. This factorization can also be re-used when solving the enriched global problem in the GFEM^{gl} [10]. An iterative algorithm for the standard FEM can be used to solve a system of equations with coefficients given by $\boldsymbol{K}^0$ instead of a direct method. In this paper, however, a direct method is adopted.

4.1 Block Gauss-Seidel Algorithm

The Block Gauss-Seidel iterative method has been used in [15] and [20] to solve the system of equations (12). An SGFEM was adopted in these references. Rather, than factorizing $\boldsymbol{K}$, the Block Gauss-Seidel (Block GS) method factorizes the diagonal blocks $\boldsymbol{K}^0$ and $\boldsymbol{K}^{gl}$. Algorithm 1 describes the method in details.

Algorithm 1: Block GS algorithm

Input: $\boldsymbol{K}, \boldsymbol{f}, \boldsymbol{d}$
Output: $\boldsymbol{d}$
for $i = 0$ *until convergence* **do**
  $\boldsymbol{r}^{gl} \leftarrow \boldsymbol{f}^{gl} - \boldsymbol{K}^{gl,0}\boldsymbol{d}^0$
  $\boldsymbol{d}^{gl} \leftarrow (\boldsymbol{K}^{gl})^{-1}\boldsymbol{r}^{gl}$
  $\boldsymbol{r}^0 \leftarrow \boldsymbol{f}^0 - \boldsymbol{K}^{0,gl}\boldsymbol{d}^{gl}$
  $\boldsymbol{d}^0 \leftarrow (\boldsymbol{K}^0)^{-1}\boldsymbol{r}^0$
return $\boldsymbol{d}$

4.2 *Preconditioned Conjugate Gradient Method*

The Preconditioned Conjugate Gradient (PCG) method is one of the most used iterative methods to solve symmetric positive-definite systems of equations. An excellent introduction to PCG can be found in [35]. The method finds new search directions through A-orthogonalization of previous search directions. It finds the magnitude of this direction by using the residual, a preconditioner, and matrix $\boldsymbol{K}$. The PCG algorithm as described in [35] follows in Algorithm 2.

The effectiveness of the PCG depends on the symmetric positive definite preconditioner $\boldsymbol{M}$ adopted. Matrix $\boldsymbol{M}$ is usually similar to $\boldsymbol{K}$ but easier to factorize. The lower the condition number of $\boldsymbol{M}^{-1}\boldsymbol{K}$, the faster the convergence of PCG. The most effective preconditioner is one that is easy to compute and factorize while leading to a better condition number than $\boldsymbol{K}$. The Block Jacobi and the Block Gauss-Seidel preconditioners are adopted in this paper. They are briefly described next.

4.2.1 Block Jacobi Preconditioner

The Block Jacobi PCG is used in [23] to solve the systems of equations (12). The following preconditioner is adopted in this algorithm

$$\boldsymbol{M} = \begin{pmatrix} \boldsymbol{K}^0 & \boldsymbol{0} \\ \boldsymbol{0} & \boldsymbol{K}^{gl} \end{pmatrix}. \tag{13}$$

This is potentially a good preconditioner for the SGFEM and SGFEM^{gl} since in these methods $\boldsymbol{K}^{gl}$ and $\boldsymbol{K}^0$ are near orthogonal [1]. The Block Jacobi Preconditioner proceeds as described in Algorithm 3.

Algorithm 2: PCG algorithm

Input: $\boldsymbol{K}, \boldsymbol{f}, \boldsymbol{d}$
Output: $\boldsymbol{d}$
$i \Leftarrow 0$
$\boldsymbol{r} \Leftarrow \boldsymbol{f} - \boldsymbol{K}\boldsymbol{u}$
$\boldsymbol{d} \Leftarrow \boldsymbol{M}^{-1}\boldsymbol{r}$
$\delta_{new} \Leftarrow \boldsymbol{r}^T\boldsymbol{d}$
for $i = 0$ *until convergence* **do**
  $\boldsymbol{q} \Leftarrow \boldsymbol{K}\boldsymbol{d}$
  $\alpha \Leftarrow \frac{\delta_{new}}{\boldsymbol{d}^T\boldsymbol{q}}$
  $\boldsymbol{d} \Leftarrow \boldsymbol{d} + \alpha\boldsymbol{d}$
  if *i is divisible by 50* **then**
    $\boldsymbol{r} \Leftarrow \boldsymbol{f} - \boldsymbol{K}\boldsymbol{u}$ /* Reset $\boldsymbol{r}$ to exact value */
  else
    $\boldsymbol{r} \Leftarrow \boldsymbol{r} - \alpha\boldsymbol{q}$ /* $\boldsymbol{r}$ is typically not evaluated directly to save computations */
  $\boldsymbol{s} \Leftarrow \boldsymbol{M}^{-1}\boldsymbol{r}$ /* Preconditioner solution step */
  $\delta_{old} \Leftarrow \delta_{new}$
  $\delta_{new} \Leftarrow \boldsymbol{r}^T\boldsymbol{s}$
  $\eta \Leftarrow \frac{\delta_{new}}{\delta_{old}}$
  $\boldsymbol{d} \Leftarrow \boldsymbol{s} + \eta\boldsymbol{d}$
return $\boldsymbol{d}$

Algorithm 3: BJ preconditioner algorithm

Input: $\boldsymbol{K}, \boldsymbol{r}$
Output: $\boldsymbol{s}$
$\boldsymbol{s}^{gl} \leftarrow (\boldsymbol{K}^{gl})^{-1}\boldsymbol{r}^{gl}$
$\boldsymbol{s}^{0} \leftarrow (\boldsymbol{K}^{0})^{-1}\boldsymbol{r}^{0}$
return $\boldsymbol{s}$

In this paper, the Cholesky factorization of $\boldsymbol{K}^0$ and $\boldsymbol{K}^{gl}$ are computed and stored at the start of the PCG iteration. See also Remark 2.

4.2.2 Block Gauss-Seidel Preconditioner

The Block Gauss-Seidel PCG uses the Block Gauss-Seidel Algorithm 4 as $\boldsymbol{M}$. The BGS method is similar computationally to the Block Jacobi method. It solves two systems of equations using the factorizations of $\boldsymbol{K}^0$ and $\boldsymbol{K}^{gl}$.

BGS-PCG is different from BJ-PCG because it involves sparse matrix multiplication of $\boldsymbol{K}^{0,gl}$ and its transpose, which is relatively inexpensive. This means that

Algorithm 4: Block GS algorithm

Input: $\boldsymbol{K}, \boldsymbol{r}, \boldsymbol{s}$
Output: $\boldsymbol{s}$
$\boldsymbol{s}^{gl} \leftarrow (\boldsymbol{K}^{gl})^{-1}\left(\boldsymbol{r}^{gl} - \boldsymbol{K}^{gl,0}\boldsymbol{s}^{0}\right)$
$\boldsymbol{s}^{0} \leftarrow (\boldsymbol{K}^{0})^{-1}\left(\boldsymbol{r}^{0} - \boldsymbol{K}^{0,gl}\boldsymbol{s}^{gl}\right)$
return $\boldsymbol{s}$

BGS-PCG considers more information about the coupling matrices. The BGS-PCG can also perform multiple iterations. In this paper, one iteration of the BGS is used in the preconditioner step of the PCG iteration. The initial value for $\boldsymbol{r}$ is $\boldsymbol{0}$.

5 Analysis of a 3-D Edge-Crack

The 3-D edge-crack shown in Fig. 2 is analyzed in this section using the GFEM^{gl} and the SGFEM^{gl}. The system of equations is solved using the iterative algorithms described in the previous section. The problem domain and dimensions are shown in Fig. 2. The dimensions are $b = 2$, $l = 4$, $t = 1$, and $a = 1$. The magnitude of the tractions is taken as $\sigma = 1$. The material parameters are Young's modulus $E = 200,000$, and Poisson's ratio $\nu = 0.3$. Point displacement boundary conditions are assigned to selected nodes of the FEM mesh to prevent rigid body motions. The GFEM^{gl} and the SGFEM^{gl} are used to numerically define the enrichment functions adopted in the global problem. The three steps of the (S)GFEM^{gl} analysis of this problem are illustrated in Fig. 3. It shows the domains for each stage in the GFEM^{gl} and SGFEM^{gl} solution process. Tetrahedron elements are used at both global and local problems. The local step of the GFEM^{gl} simulates the crack. Spring boundary

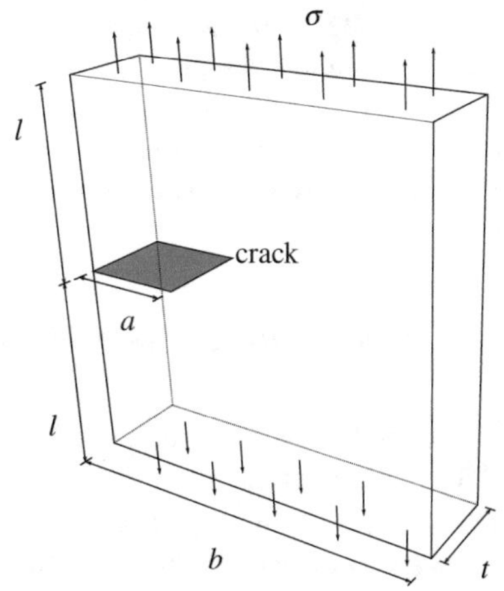

Fig. 2 Three-dimensional edge crack

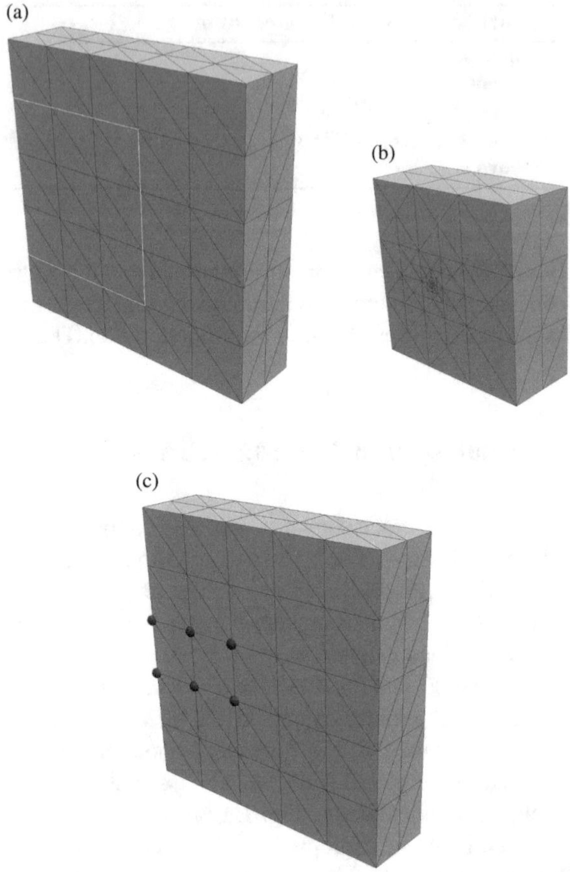

Fig. 3 GFEM^{gl} steps for a 3-D edge crack. The same steps are used in the SGFEM^{gl}. Red spheres are shown at global nodes enriched with the local solution. (**a**) Initial global mesh. (**b**) Local mesh. (**c**) Enriched global mesh

conditions are applied along the portion of the local boundary that does not intersect the boundary of the global problem. The local mesh is refined near the crack front, with the element length adjacent to the crack front being about 5% of the crack length. The polynomial order of the local problem is taken as 3. The local solution is used to generate enrichments for the enriched global problem. These global-local enrichments are the only enrichments in the global domain.

5.1 Condition Number Analysis

The condition number of the global stiffness matrix $\boldsymbol{K}$ of the GFEM^{gl} and the SGFEM^{gl} is compared in this section. The condition number of the sub-matrices in (12) is also compared. The following notation is adopted hereafter: $\kappa(\boldsymbol{K})$, $\kappa(\boldsymbol{K}^0)$, and $\kappa(\boldsymbol{K}^{gl})$ denote the condition number of global matrix $\boldsymbol{K}$, and sub-matrices $\boldsymbol{K}^0$

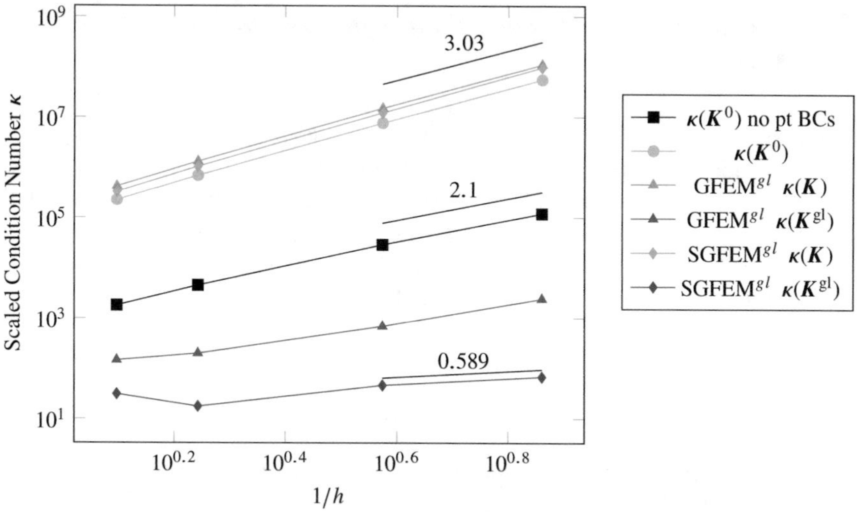

Fig. 4 Growth of condition numbers as the global mesh is refined in all three directions. Point Dirichlet boundary conditions are prescribed to the global problem to prevent rigid body motion

and $\boldsymbol{K}^{\mathrm{gl}}$, respectively. The condition number is computed for a sequence of global meshes with the same number of elements in all three directions. The local mesh is unchanged. This, and the boundary condition of the global problem, implies that the global-local enrichments do not change with refinement of the global mesh.

Figure 4 shows the condition number for the GFEM^{gl} and the SGFEM^{gl} with this sequence of meshes. The plots show that $\kappa(\boldsymbol{K}) = O(h^{-3})$ for both methods and that $\kappa(\boldsymbol{K}^0)$ is also of $O(h^{-3})$. This is surprising since one would expect that the conditioning of $\boldsymbol{K}$ would grow much quicker than $\kappa(\boldsymbol{K}^0)$, at least for the GFEM^{gl}. The cause of this apparent contradiction is the point Dirichlet boundary conditions prescribed to prevent rigid body motion of the global problem. The condition number of the FEM stiffness matrix for a 3-D Neumann problem with point boundary conditions is $O(h^{-3})$ [8]. Thus, the condition number for both the GFEM^{gl} and the SGFEM^{gl} matrices is controlled by the effect of point constraints. Figure 4 also shows $\kappa(\boldsymbol{K}^0)$ when no point constraint is prescribed to the global problem. In this case $\kappa(\boldsymbol{K}^0) = O(h^{-2})$ as expected. It is noted that in the case of the GFEM^{gl}, $\kappa(\boldsymbol{K})$ is expected to grow faster than $O(h^{-3})$ with further mesh refinement than shown in Fig. 4. This is proved in [1].

Another interesting feature shown in Fig. 4 is the noticeable decrease in SGFEM^{gl} $\kappa(\boldsymbol{K}^{\mathrm{gl}})$ relative to $\kappa(\boldsymbol{K}^{\mathrm{gl}})$ from GFEM^{gl}. This is the case even though the SGFEM was designed to reduce the condition number $\kappa(\boldsymbol{K})$ by reducing the coupling between $\boldsymbol{K}^0$ and $\boldsymbol{K}^{\mathrm{gl}}$. This reduction in $\kappa(\boldsymbol{K}^{\mathrm{gl}})$ has an impact on the performance of iterative solvers as shown in the next section.

5.2 Performance of Preconditioners for GFEM^{gl} and SGFEM^{gl}

The performance of the iterative algorithms described in Sect. 4 is investigated in this section. The CG, BJ-PCG and BGS-PCG algorithms are used to solve the global system of equations (12) associated with the GFEM^{gl} and the SGFEM^{gl} discretizations. Each iterative solver is run until the relative error is less than $e^{\text{conv}} = 10^{-5}$, which is taken as the convergence tolerance. The relative error e^i of an iterative solver solution is calculated at each iteration i using

$$e^i = \frac{||\hat{\boldsymbol{d}} - \boldsymbol{d}^i||_2}{||\hat{\boldsymbol{d}}||_2}$$

where $\hat{\boldsymbol{d}}$ is a precalculated direct solver solution. The iteration i at which the solver converges is hereafter denoted i^{conv}.

Figure 5 shows the number of iterations for convergence (i^{conv}) of each solver and for the GFEM^{gl} and SGFEM^{gl}. The same sequence of global meshes adopted in the previous section is used. Point Dirichlet boundary conditions are used in the global problem to prevent rigid body motion. For any given mesh, the GFEM^{gl} i^{conv} is significantly higher than that for the SGFEM^{gl}. Also, the rate of growth of i^{conv} with respect to element size for the SGFEM^{gl} is less than half the rate for the GFEM^{gl}. This indicates that although the conditioning of the matrices for GFEM^{gl} and SGFEM^{gl} are for this problem fairly similar, iteratively solving (12) for the SGFEM^{gl} is faster than for the GFEM^{gl}. Both BGS-PCG and BJ-PCG benefit from the SGFEM^{gl}. The slopes of the curves in Fig. 5 is similar but i^{conv} is always less for the BGS-PCG than for BJ-PCG. The advantages of SGFEM versus GFEM in iterative solvers are also shown in [15] and [20].

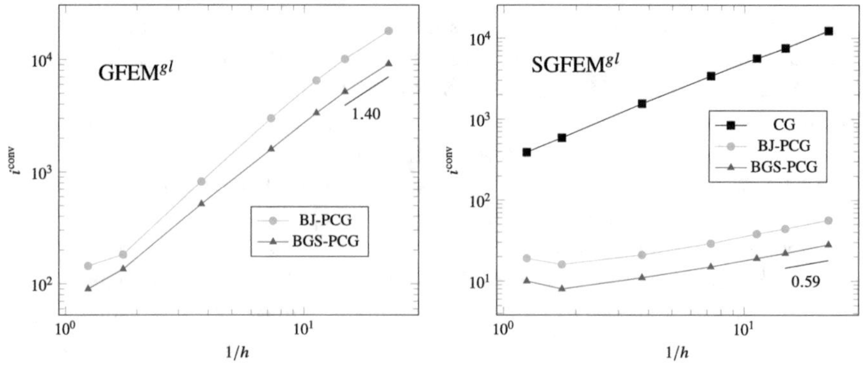

Fig. 5 Iterations to convergence of CG, BJ-PCG and BGS-PCG when solving global system (12) for GFEM^{gl} (left) and SGFEM^{gl} (right). Point Dirichlet boundary conditions are adopted in the global problem

5.3 Comparison of BGS-PCG with Pardiso

The performance of the proposed BGS-PCG algorithm with the SGFEM^{gl} is compared with the Intel Pardiso direct solver [24, 33]. The pure Neumann edge-crack problem with point Dirichlet boundary conditions is solved using both solvers and the sequence of uniform meshes described earlier. The largest global problem has about 2 million degrees of freedom. The CPU time required for convergence of the BGS-PCG when solving (12) and for the factorization of $\boldsymbol{K}$ by Pardiso is plotted against the number of degrees of freedom in Fig. 6. It shows that the BGS-PCG is, for this problem, always faster than Pardiso. The slope of the BGS-PCG curve is lower than the one for Pardiso which implies that the bigger the problem, the more efficient the BGS-PCG is relative to Pardiso. For the largest problem solved, BGS-PCG took 496 s for convergence while Pardiso required 7030 s for the factorization of $\boldsymbol{K}$ which is about 14 times slower than BGS-PCG.

The slope of 2.07 for Pardiso and 1.54 for BGS-PCG shown in Fig. 6 can be considered equivalent to the rate of increase of the total number of algebraic operations versus the number of degrees of freedom of the problem. The theoretical rate when solving 3-D elliptic boundary value problems using the adopted sparse direct solver is 2 [19]. This matches pretty well with the rate shown in Fig. 6. The theoretical rate when solving the same class of problems using the preconditioned conjugate gradient is 1.17 [19], which is lower than the rate for the BGS-PCG shown in Fig. 6. This can be traced to the computational effort required by the BGS preconditioner. The cost of the PCG is given by the number of CG iterations times the cost of each iteration, including the preconditioner. Figure 5 shows that the number of iterations required for convergence of the BGS-PCG grows at a rate of 0.59/3 with respect to the number of degrees of freedom in 3-D. This slow rate

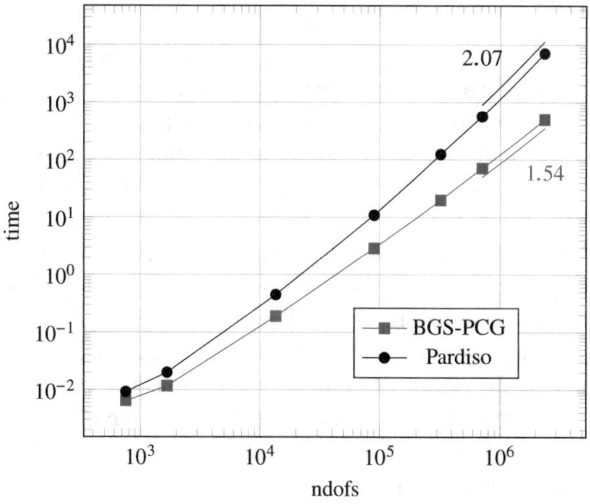

Fig. 6 Comparison of BGS-PCG with Pardiso for several problem sizes

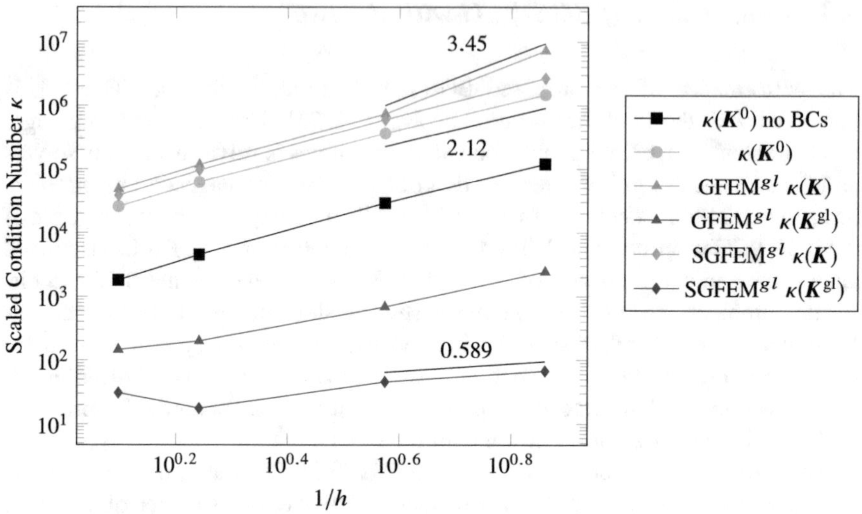

Fig. 7 Growth of condition numbers as the global mesh is refined in all three directions. Face Dirichlet boundary conditions are prescribed at the bottom face of the global domain instead of point constraints

of growth indicates that the higher-than theoretical rate observed in Fig. 6 for the BGS-PCG is likely caused by the cost of the BGS preconditioner.

5.4 Condition Number with Face Dirichlet Boundary Conditions

The boundary conditions applied at the lower edge of the domain shown in Fig. 2 are in this section changed from a constant traction to prescribed displacements in all directions. This eliminates the need to impose point constraints to the body. The effect of this change on the condition number of the matrices is shown in Fig. 7. Two changes can be observed relative to the results shown in Fig. 4. First, the growth rate of $\kappa(\boldsymbol{K})$ of the SGFEM^{gl} and $\kappa(\boldsymbol{K}^0)$ are close to the growth rate of $\kappa(\boldsymbol{K}^0)$ without boundary conditions. Recall that $\boldsymbol{K}^0$ is a standard FEM matrix. Second, the growth rate of $\kappa(\boldsymbol{K})$ of the GFEM^{gl} becomes higher than in the SGFEM^{gl} as the mesh is refined, attesting the benefits of using the SGFEM^{gl}.

The performance of the BJ-PCG and BGS-PCG iterative solvers with the change in boundary conditions described above is nearly identical to the Neumann problem with point constraints. This is due the choice of $\boldsymbol{K}^0$ as a preconditioner. The point constraints are applied to this matrix, and thus its effect on the CG is absorbed by the preconditioner.

6 Conclusions

The main conclusions based on the numerical experiments presented in Sect. 5 are summarized in this section.

The convergence of iterative solvers for the GFEM^{gl} and SGFEM^{gl} is controlled by the element size adopted in the global problem. This is important since, for the same level of accuracy, these methods can adopt much coarser meshes than the FEM [16]. The mesh size in the FEM has to be comparable to the local problem mesh size within the GFEM^{gl} and SGFEM^{gl}.

The SGFEM modification of enrichment functions can also be applied to numerically computed enrichments leading to the SGFEM^{gl}. This was also shown in [15]. The results presented in Sect. 5 show that: (1) The SGFEM^{gl} requires much fewer PCG iterations for convergence than the GFEM^{gl}; (2) The proposed preconditioners significantly accelerate the convergence of the SGFEM^{gl}. They also reduce the rate of growth of the number of iterations for convergence with respect to problem size.

Acknowledgements T.B. Fillmore and C.A. Duarte gratefully acknowledge the research funding under contract number AF Sub OSU 60038238 provided by the Collaborative Center in Structural Sciences (C^2S^2) at the Ohio State University, supported by the U.S. Air Force Research Laboratory.

References

1. I. Babuška, U. Banerjee, Stable generalized finite element method (SGFEM). Tech. Report ICES REPORT 11–07, The Institute for Computational Engineering and Sciences, The University of Texas at Austin, April 2011
2. I. Babuška, U. Banerjee, Stable generalized finite element method (SGFEM). Comput. Methods Appl. Mech. Eng. **201–204**, 91–111 (2012)
3. I. Babuška, J.M. Melenk, The partition of unity method. Int. J. Numer. Methods Eng. **40**, 727–758 (1997)
4. I. Babuška, G. Caloz, J.E. Osborn, Special finite element methods for a class of second order elliptic problems with rough coefficients. SIAM J. Numer. Anal. **31**(4), 945–981 (1994)
5. E. Béchet, H. Minnebo, N. Moës, B. Burgardt, Improved implementation and robustness study of the x-fem for stress analysis around cracks. Int. J. Numer. Methods Eng. **64**, 1033–1056 (2005)
6. T. Belytschko, T. Black, Elastic crack growth in finite elements with minimal remeshing. Int. J. Numer. Methods Eng. **45**, 601–620 (1999)
7. L. Berger-Vergiat, H. Waisman, B. Hiriyur, R. Tuminaro, D. Keyes, Inexact Schwarz-AMG preconditioners for crack problems modeled by XFEM. Int. J. Numer. Methods Eng. **90**, 311–328 (2012)
8. P. Bochev, R.B. Lehoucq, On the finite element solution of the pure Neumann problem. SIAM Rev. **47**(1), 50–66 (2005)
9. A.Th. Diamantoudis, G.N. Labeas, Stress intensity factors of semi-elliptical surface cracks in pressure vessels by global-local finite element methodology. Eng. Fract. Mech. **72**, 1299–1312 (2005)

10. C.A. Duarte, D.-J. Kim, Analysis and applications of a generalized finite element method with global-local enrichment functions. Comput. Methods Appl. Mech. Eng. **197**(6–8), 487–504 (2008)
11. C.A.M. Duarte, J.T. Oden, An hp adaptive method using clouds. Comput. Methods Appl. Mech. Eng. **139**, 237–262 (1996)
12. C.A. Duarte, I. Babuška, J.T. Oden, Generalized finite element methods for three dimensional structural mechanics problems. Comput. Struct. **77**, 215–232 (2000)
13. C.A. Duarte, O.N. Hamzeh, T.J. Liszka, W.W. Tworzydlo, A generalized finite element method for the simulation of three-dimensional dynamic crack propagation. Comput. Methods Appl. Mech. Eng. **190**(15–17), 2227–2262 (2001)
14. C.A. Duarte, D.-J. Kim, I. Babuška, A global-local approach for the construction of enrichment functions for the generalized FEM and its application to three-dimensional cracks, in *Advances in Meshfree Techniques*, ed. by V.M.A. Leitão, C.J.S. Alves, C.A. Duarte. Computational Methods in Applied Sciences, vol. 5 (Springer, Dordrecht, 2007), pp. 1–26
15. V. Gupta, Improved conditioning and accuracy of a two-scale generalized finite element method for fracture mechanics. Ph.D. thesis, University of Illinois at Urbana-Champaign, 2014
16. V. Gupta, D.-J. Kim, C.A. Duarte, Analysis and improvements of global-local enrichments for the generalized finite element method. Comput. Methods Appl. Mech. Eng. **245–246**, 47–62 (2012)
17. V. Gupta, C.A. Duarte, I. Babuška, U. Banerjee, A stable and optimally convergent generalized FEM (SGFEM) for linear elastic fracture mechanics. Comput. Methods Appl. Mech. Eng. **266**, 23–39 (2013)
18. V. Gupta, C.A. Duarte, I. Babuška, U. Banerjee, Stable GFEM (SGFEM): improved conditioning and accuracy of GFEM/XFEM for three-dimensional fracture mechanics. Comput. Methods Appl. Mech. Eng. **289**, 355–386 (2015)
19. M.T. Heath, *Scientific Computing: An Introductory Survey*, 2nd edn. McGraw-Hill Series in Computer Science (McGraw-Hill, Boston, MA, 2002)
20. K. Kergrene, I. Babuska, U. Banerjee, Stable generalized finite element method and associated iterative schemes; application to interface problems. Comput. Methods Appl. Mech. Eng. **305**, 1–36 (2016)
21. D.-J. Kim, C.A. Duarte, J.P. Pereira, Analysis of interacting cracks using the generalized finite element method with global-local enrichment functions. J. Appl. Mech. **75**(5), 1–12 (2008)
22. D.-J. Kim, J.P. Pereira, C.A. Duarte, Analysis of three-dimensional fracture mechanics problems: a two-scale approach using coarse generalized FEM meshes. Int. J. Numer. Methods Eng. **81**(3), 335–365 (2010)
23. D.-J. Kim, S.-G. Hong, C.A. Duarte, Generalized finite element analysis using the preconditioned conjugate gradient method. Appl. Math. Modell. **39**(19), 5837–5848 (2015)
24. A. Kuzmin, M. Luisier, O. Schenk, Fast methods for computing selected elements of the greens function in massively parallel nanoelectronic device simulations. Euro-Par 2013 Parallel Process. **8097**, 533–544 (2013)
25. M. Malekan, F.B. Barros, Well-conditioning global-local analysis using stable generalized/extended finite element method for linear elastic fracture mechanics. Comput. Mech. **58**(5), 819–831 (2016)
26. J.M. Melenk, I. Babuška, The partition of unity finite element method: basic theory and applications. Comput. Methods Appl. Mech. Eng. **139**, 289–314 (1996)
27. A. Menk, S. Bordas, A robust preconditioning technique for the extended finite element method. Int. J. Numer. Methods Eng. **85**(13), 1609–1632 (2011)
28. N. Moës, J. Dolbow, T. Belytschko, A finite element method for crack growth without remeshing. Int. J. Numer. Methods Eng. **46**, 131–150 (1999)
29. A.K. Noor, Global-local methodologies and their applications to nonlinear analysis. Finite Elem. Anal. Des. **2**, 333–346 (1986)

30. J.T. Oden, C.A. Duarte, Clouds, cracks and FEMs, in *Recent Developments in Computational and Applied Mechanics*, ed. by B.D. Reddy (International Center for Numerical Methods in Engineering, CIMNE, Barcelona, 1997), pp. 302–321. http://gfem.cee.illinois.edu/papers/jMartin_color.pdf
31. J.T. Oden, C.A. Duarte, O.C. Zienkiewicz, A new cloud-based *hp* finite element method. Comput. Methods Appl. Mech. Eng. **153**, 117–126 (1998)
32. J.P. Pereira, D.-J. Kim, C.A. Duarte, A two-scale approach for the analysis of propagating three-dimensional fractures. Comput. Mech. **49**(1), 99–121 (2012)
33. O. Schenk, K. Gärtner, Solving unsymmetric sparse systems of linear equations with PARDISO. J. Futur. Gener. Comput. Syst. **20**(9), 475–487 (2004)
34. M.A. Schweitzer, Generalizations of the finite element method. Cen. Eur. J. Math. **10**, 3–24 (2012)
35. J.R. Shewchuk, An introduction to the conjugate gradient method without the agonizing pain (1994)
36. H. Waisman, L. Berger-Vergiat, An adaptive domain decomposition preconditioner for crack propagation problems modeled by XFEM. Int. J. Multiscale Comput. Eng. **11**(6), 633–654 (2013)

A Fast and Stable Multi-Level Solution Technique for the Method of Fundamental Solutions

Csaba Gáspár

Abstract The classical form of the Method of Fundamental Solutions is applied. Instead of using a single set of subtly located external sources, a special strategy of defining several sets of external source points is introduced. The sets of sources are defined by the quadtree/octtree subdivision technique controlled by the boundary collocation points in a completely automatic way, resulting in a point set, the density of the spatial distribution of which decreases quickly far from the boundary. The 'far' sources are interpreted to form a 'coarse grid', while the densely distributed 'near-boundary' sources are considered a 'fine grid' (despite they need not to have any grid structure). Based on this classification, a multi-level technique is built up, where the smoothing procedure is defined by performing some familiar iterative technique e.g. the (conjugate) gradient method. The approximate solutions are calculated by enforcing the boundary conditions in the sense of least squares. The resulting multi-level method is robust and significantly reduces the computational cost. No weakly or strongly singular integrals have to be evaluated. Moreover, the problem of severely ill-conditioned matrices is completely avoided.

1 Introduction

The Method of Fundamental Solutions (MFS, see. e.g. [10]) has become a popular, truly meshfree solution technique of some elliptic boundary value problems due to its simplicity and high accuracy. In addition to it, the MFS is a boundary-only technique, that is, it requires some (unstructured) points along the boundary but not inside the domain.

C. Gáspár (✉)
Széchenyi István University, Győr, Hungary
e-mail: gasparcs@sze.hu

M. Griebel, M. A. Schweitzer (eds.), *Meshfree Methods for Partial Differential Equations IX*, Lecture Notes in Computational Science and Engineering 129,
https://doi.org/10.1007/978-3-030-15119-5_2

If L is a 2D or 3D elliptic partial differential operator, Ω is a (2D or 3D) bounded domain with boundary Γ, consider the boundary value problem

$$Lu = 0 \quad \text{in } \Omega \tag{1}$$

$$u|_\Gamma = u_0, \tag{2}$$

where a simple Dirichlet boundary condition is supposed for simplicity. Denote by Φ a fundamental solution of L, then the traditional form of the MFS provides an approximate solution of (1)–(2) in the following form:

$$u(x) = \sum_{j=1}^{N} \alpha_j \Phi(x - \tilde{x}_j) \tag{3}$$

Here $\tilde{x}_1, \tilde{x}_2, \ldots, \tilde{x}_N$ are predefined external points (*source points*). Thus, the partial differential equation (1) is exactly satisfied in Ω; due to the singularity of Φ at the origin, the approximate solution u defined by (3) exhibits singularities at the source points (but not in Ω).

The a priori unknown coefficients $\alpha_1, \alpha_2, \ldots, \alpha_N$ are determined by enforcing the boundary conditions in some *boundary collocation points* $x_1, x_2, \ldots, x_M \in \Gamma$, i.e. by solving the linear system of equations:

$$\sum_{j=1}^{N} \alpha_j \Phi(x_k - \tilde{x}_j) = u_0(x_k) =: u_k \qquad (k = 1, 2, \ldots M) \tag{4}$$

Note that the case of Neumann or mixed boundary conditions can be treated in a completely similar way. In this case, some equations of the system (4) contain the normal derivative of the fundamental solution:

$$\sum_{j=1}^{N} \alpha_j \frac{\partial \Phi}{\partial n_k}(x_k - \tilde{x}_j) = v_0(x_k) =: v_k \qquad (\text{if } x_k \in \Gamma_N)$$

Here n_k is the outward normal unit vector at the point x_k and Γ_N denotes the Neumann part of the boundary Γ, where the boundary condition

$$\frac{\partial u}{\partial n}|_{\Gamma_N} = v_0$$

is prescribed.

The matrix of the linear system (4) is generally non-selfadjoint and fully populated. The numbers of the source points and boundary collocation points need not be equal. In this case, the linear system (4) can be dealt with either the Singular Value Decomposition (SVD) or some iterative method applied to the Gaussian normal equations of (4) i.e. a technique based on the least squares. For technical reasons, however, it is often supposed that $M = N$, i.e. the system (4) has a square matrix. In this case, the system (4) is often severely ill-conditioned, which causes numerical difficulties, though the accuracy of the method is generally very good, see [12]. This is especially the case, when the sources are located far from the boundary. On the other hand, if they are too close to the boundary, numerical singularities appear in the vicinity of the boundary collocation points, which may strongly reduce the accuracy.

Another problem in implementing the MFS is the proper and automated definition of the external source points. As pointed out by many authors, this is not a trivial task, especially in such cases when the shape of the domain Ω is complicated. Moreover, both the computational complexity and the accuracy are sensitive to the choice of the source locations. In [1, 5] the use of sources with large distance from the boundary was proposed. This results in excellent accuracy (provided that the exact solution is smooth enough), but leads to an extremely ill-conditioned linear system to be solved. In [16], the number of sources may be initially quite large, and several algorithms were proposed to pick out a (preferably much) smaller amount of sources which results in approximately the same accuracy. More recently, in [4], several strategies are investigated to define source locations, but all of them were defined to be placed along the boundary of a larger domain.

To avoid the above mentioned difficulties, a number of techniques have been developed. A group of such methods is based on allowing the source and the boundary collocation points to coincide. This solves the problem of the automatic location of sources. Using nonsingular solutions instead of the fundamental solution, the problem of singularity can be avoided [3], but the problem of ill-conditioned matrices remains. The picture is similar, if fundamental solutions concentrated to straight lines are used [7].

The use of the classical fundamental solutions leads to the problem of proper computations of singular terms, since the diagonal entries of the matrix of (4) have to be properly redefined due to the singularity of Φ. The problem is more difficult in the presence of Neumann or mixed boundary conditions, since the normal derivative of Φ has a stronger singularity at the origin. These methods are some regularization and/or more sophisticated desingularization techniques (see [14, 15, 18]), often based on the solution of some auxiliary problem (see [2, 8, 11]).

In this paper, we return to the traditional form of the MFS, using several groups of source points. The spatial density of the source points decreases rapidly when their distance from the boundary increases. Such point sets can be easily defined by using the computationally very efficient quadtree/octtree subdivision technique. The subdivision is controlled by the boundary collocation points and can be performed in a completely automatic way. Using the multigrid terminology, the 'far' sources are interpreted to form a 'coarse grid', while the densely distributed 'near-boundary' sources are considered a 'fine grid'. It should be pointed out, however, that they need not to have any grid structure. On each level, we use much more boundary collocation points than source points. As a smoothing procedure, a simple (conjugate) gradient method is applied to the Gaussian normal equations of the corresponding system having the form (4). This significantly reduces the high-frequency components of the error. Using the familiar multigrid tools, a simple multi-level algorithm is constructed resulting in a computationally economic method, which still has an acceptable accuracy. As the main advantage, it should be pointed out that the sources are defined in an automatic way, and, at the same time, the problem of highly ill-conditioned systems is completely avoided.

2 The Method of Fundamental Solutions with Near-Boundary Sources

For the sake of simplicity, we restrict ourselves to the 2D Laplace equation supplied with pure Dirichlet boundary condition. Suppose that the domain of the partial differential equation is a circle centered at the origin with radius R: $\Omega_R := \{(x, y) \in \mathbf{R}^2 : x^2 + y^2 < R^2\}$. The problem to be investigated is

$$\Delta u = 0 \quad \text{in } \Omega_R, \tag{5}$$

$$u|_{\Gamma_R} = u_0, \tag{6}$$

where Γ_R is the boundary of Ω_R and u_0 is a predefined boundary function. It is well known that if u_0 is regular enough, problem (5)–(6) has a unique solution in an appropriate Sobolev space.

Let us seek the exact solution in the form of a single layer potential (the discretized form of this technique leads to the traditional Method of Fundamental Solutions). Let the sources of the single layer potential be concentrated on the circle $\Gamma_{R+\delta}$ centered at the origin with radius $R + \delta$, where $\delta > 0$ is a given constant:

$$(\Phi w)(x) := \int_{\Gamma_{R+\delta}} (\log ||x - y||) \cdot w(y)\, d\Gamma_y \tag{7}$$

($x \in \overline{\Omega}_{R+\delta}$; $||.||$ denotes the Euclidean norm in $\mathbf{R}^2$). Suppose that the density function w is expressed in terms of complex Fourier series (in polar coordinates):

$$w(R + \delta, t) = \sum_k w_k \cdot e^{ikt} \tag{8}$$

where, for simplicity, we assume that $w_0 = \frac{1}{2\pi} \int_{-\pi}^{\pi} w(t)\, dt = 0$. Standard calculations show that the single layer potential Φw along the circle $\Gamma_{R+\delta}$ is as follows (written in polar coordinates):

$$(\Phi w)(R + \delta, t) = -(R + \delta)\pi \cdot \sum_{k \neq 0} \frac{1}{|k|} w_k \cdot e^{ikt}. \tag{9}$$

Moreover, for the point $(r, t) \in \overline{\Omega}_{R+\delta}$, the single layer potential can be expressed as:

$$(\Phi w)(r,t) = -(R+\delta)\pi \sum_{k \neq 0} \frac{1}{|k|} w_k \left(\frac{r}{R+\delta} \right)^{|k|} \cdot e^{ikt}. \tag{10}$$

(Indeed, this function is harmonic inside $\overline{\Omega}_{R+\delta}$ and identically equals (9) along the boundary $\Gamma_{R+\delta}$.)

Let us express the boundary condition u_0 in terms of complex Fourier series as well:

$$u_0(t) := \sum_{k \neq 0} \beta_k \cdot e^{ikt} \tag{11}$$

Then, from (11) and (10), we immediately obtain that the Fourier coefficients of the density function w can be expressed with those of the boundary condition u_0 as

$$w_k = -\frac{|k|}{(R+\delta)\pi} \left(1 + \frac{\delta}{R} \right)^{|k|} \cdot \beta_k. \tag{12}$$

Define the operator

$$Aw := (\Phi w)|_{\Gamma_R} = -(R+\delta)\pi \cdot \sum_{k \neq 0} \frac{1}{|k|} w_k \left(\frac{R}{R+\delta} \right)^{|k|} \cdot e^{ikt}. \tag{13}$$

We have obtained that the (unique) solution of the equation

$$Aw = u_0 \tag{14}$$

defines the (unique) solution of (5)–(6) in the form:

$$u = \Phi w \tag{15}$$

where $w = A^{-1}u_0$, and the Fourier coefficients of w can be computed by (12).

Note that the operator A^{-1} is not bounded between the usual Sobolev spaces $H^{s_1}(\Gamma_R)$ and $H^{s_2}(\Gamma_{R+\delta})$ due to the exponentially increasing factor $\left(1 + \frac{\delta}{R}\right)^{|k|}$, which may cause serious numerical difficulties. However, the inverses of the discretizations of A can still be uniformly bounded, if the fineness of the discretization depends on the distance δ. Roughly speaking, if the discretization of A becomes finer and finer, the distance δ of Γ_R and $\Gamma_{R+\delta}$ should be smaller and smaller. Two of such techniques are detailed below.

2.1 Band-Limited Approximation

Let $N \in \mathbf{N}$ be a fixed even index (a power of 2 in the following). Let us look for an approximate solution of (14) in terms of truncated Fourier series, i.e.

$$w^{(N)}(t) := \sum_{0 \neq |k| \leq N/2} w_k \cdot e^{ikt}. \tag{16}$$

Denote by X_N the N-dimensional subspace of the functions in $L_2(0, 2\pi)$ spanned by the functions e^{ikt} $(0 \neq |k| \leq N/2)$, and define the operator $A^{(N)}$ by restricting A to X_N, i.e.

$$(A^{(N)} w^{(N)})(t) := -(R+\delta)\pi \cdot \sum_{0 \neq |k| \leq N/2} \frac{1}{|k|} w_k \left(\frac{R}{R+\delta}\right)^{|k|} \cdot e^{ikt}.$$

Then the equation

$$A^{(N)} w^{(N)} = u_0 \tag{17}$$

has no solution in general but has a unique generalized solution in the sense of least squares. Namely, define the Fourier coefficients w_k by (12) for $0 \neq |k| \leq N/2$, while for all $|k| > N/2$, define $w_k := 0$. This function $w^{(N)}$ minimizes $||A^{(N)} w - u_0||^2_{L_2(0,2\pi)}$ on X_N, since by Parseval's formula

$$||A^{(N)} w^{(N)} - u_0||^2_{L_2(0,2\pi)} =$$

$$= 2\pi \cdot \sum_{0 \neq |k| \leq N/2} \left| -\frac{(R+\delta)\pi}{|k|} \left(\frac{R}{R+\delta}\right)^{|k|} w_k - \beta_k \right|^2 + 2\pi \cdot \sum_{|k| > N/2} |\beta_k|^2.$$

The first sum of the right-hand side equals to zero due to the definition of w_k (see (12)), while the second sum is independent of $w^{(N)}$. That is, the above function $w^{(N)}$ minimizes the L_2-norm of the residual $(A^{(N)} w - u_0)$.

2.1.1 Accuracy

The accuracy of the band-limited approximate solution can be characterized by some norm of the residual $(A^{(N)} w^{(N)} - u_0)$. Let $s \geq 1$ be arbitrary, then

$$||A^{(N)} w^{(N)} - u_0||^2_{L_2(0,2\pi)} = 2\pi \cdot \sum_{|k| > N/2} |\beta_k|^2 \leq$$

$$\leq \frac{2^{2s+1}\pi}{N^{2s}} \cdot \sum_{|k|>N/2} |k|^{2s}|\beta_k|^2 \leq$$

$$\leq \frac{2^{2s+1}\pi}{N^{2s}} \cdot ||u_0||^2_{H^s(\Gamma_R)},$$

which implies the following result.

Proposition 1 *If the boundary condition u_0 belongs to the Sobolev space $H^s(\Gamma_R)$ for some $s \geq 1$, then the L_2-norm of the residual of the band-limited approximation (16) can be estimated by*

$$||A^{(N)}w^{(N)} - u_0||_{L_2(0,2\pi)} \leq \frac{2^s \cdot \sqrt{2\pi}}{N^s} \cdot ||u_0||_{H^s(\Gamma_R)}. \tag{18}$$

Remark Proposition 1 states that the band-limited approximation (16) results in a quite accurate approximate solution provided that the boundary condition is smooth enough.

2.1.2 Condition Number

To numerically solve the problem (14), the most natural technique is the use of least squares. Practically, this means that the Gaussian normal equations have to be solved, i.e.

$$(A^{(N)})^* A^{(N)} w = (A^{(N)})^* u_0. \tag{19}$$

The matrix of the operator $A^{(N)}$ is diagonal in the basis of the orthogonal functions e^{ikt} $(0 \neq |k| \leq N/2)$. The diagonal entries can be obtained from (13) as

$$A^{(N)}_{kk} = -(R+\delta)\pi \cdot \frac{1}{|k|} \cdot \left(\frac{R}{R+\delta}\right)^{|k|} \qquad (0 \neq |k| \leq \frac{N}{2}),$$

which are the eigenvalues of $A^{(N)}$; the corresponding eigenfunctions are e^{ikt}. Thus, the condition number of $A^{(N)}$ can be easily calculated as

$$\text{cond}(A^{(N)}) = \frac{|A^{(N)}_{11}|}{|A^{(N)}_{N/2,N/2}|} = \frac{\frac{R}{R+\delta}}{\frac{2}{N}(\frac{R}{R+\delta})^{N/2}} \leq \frac{N}{2} \cdot \left(1 + \frac{\delta}{R}\right)^{N/2}. \tag{20}$$

It is clear that if $\delta > 0$ is fixed (independently of N), then $\text{cond}(A^{(N)})$ increases exponentially with N. However, if δ depends on N in such a way that δ is inversely proportional to N, i.e.

$$\frac{\delta}{R} = \frac{2\pi}{N} \cdot \delta_0 \tag{21}$$

with some $\delta_0 > 0$ (δ_0 is independent of N), then from (20), we have

$$\text{cond}(A^{(N)}) \leq \frac{N}{2} \cdot \left(1 + \frac{2\pi\,\delta_0}{N}\right)^{N/2} \leq \frac{N}{2} \cdot e^{\pi\,\delta_0}. \tag{22}$$

We have obtained an estimation for the condition number of the operator $A^{(N)}$.

Proposition 2 *If the distance of the boundaries Γ_R and $\Gamma_{R+\delta}$ satisfies the condition (21), then the condition number of $A^{(N)}$ can be estimated as*

$$cond(A^{(N)}) \leq \frac{e^{\pi\,\delta_0}}{2} \cdot N. \tag{23}$$

Consequently, for the condition number of the Gaussian normal equation (19), the estimation

$$cond((A^{(N)})^* A^{(N)}) \leq \frac{e^{2\pi\,\delta_0}}{4} \cdot N^2 \tag{24}$$

holds.

2.1.3 Convergence

To solve Eq. (17), some direct method e.g. Gaussian elimination is used in general. The computational cost of these methods are rather high, therefore the classical (conjugate) gradient iteration technique will be applied to the Gaussian normal equations (19). Recall (for details, see e.g. [13]) that if A is a self-adjoint, positive definite operator with condition number κ, then after m gradient steps, the error of the approximate solution of the equation $Ax = b$ can be estimated by

$$||x_m - x^*||_A \leq \left(\frac{\kappa - 1}{\kappa + 1}\right)^m \cdot ||x_0 - x^*||_A,$$

while after m conjugate gradient steps, the error estimation has the form

$$||x_m - x^*||_A \leq 2 \cdot \left(\frac{\sqrt{\kappa} - 1}{\sqrt{\kappa} + 1}\right)^m \cdot ||x_0 - x^*||_A.$$

Here x^* and x_m denotes the exact and the approximate solutions, respectively. $||.||_A$ is the energy norm, i.e. $||x||_A = \sqrt{\langle Ax, x\rangle}$.

Proposition 2 states that if the distance of the sources and the boundary Γ_R is inversely proportional to N (where N characterizes the 'fineness' of the band-limited discretization), then the condition number of the discretized operator $A^{(N)}$ increases linearly with N only. This results in moderately increasing condition numbers. However, this moderate increase makes the (conjugate) gradient method slow very soon, even for relatively low values of N. The convergence can be significantly improved by a multi-level technique detailed below.

2.1.4 Controlling by External Boundary Values

If, in contrast to the MFS-approach, the band-limited approximate solution is expressed in terms of *external boundary values on* $\Gamma_{R+\delta}$ (rather than external sources), the situation becomes simpler. Now the approximate solution has the form

$$u^{(N)}(r,t) = \sum_{|k|\le N/2} \hat{u}_k \left(\frac{r}{R+\delta}\right)^{|k|} e^{ikt}.$$

The boundary condition along Γ_R is approximated by enforcing

$$u^{(N)}(R,t) = \sum_{|k|\le N/2} \hat{u}_k \left(\frac{R}{R+\delta}\right)^{|k|} \cdot e^{ikt} = u_0(t) = \sum_k \beta_k e^{ikt}$$

in the sense of least squares, yielding

$$\hat{u}_k = \left(\frac{R+\delta}{R}\right)^{|k|} \cdot \beta_k \qquad (|k| \le \frac{N}{2}),$$

whence

$$|\beta_k| \le |\hat{u}_k| \le \left(1 + \frac{2\pi\delta_0}{N}\right)^{|k|} \cdot |\beta_k| \le e^{\pi\delta_0} \cdot |\beta_k|.$$

This implies that the computation of the a priori unknown external boundary values along $\Gamma_{R+\delta}$ is a *uniformly well-conditioned* problem; the condition number remains smaller than $e^{\pi\delta_0}$, independently of N. This means that the classical (conjugate) gradient method itself is a robust method without introducing any multi-level technique. In practice, the technique requires an external boundary and a robust method to solve the external problem. This can be performed by using quadtree-based multi-level techniques, for details, see [9]. The present method, however, requires no external boundary and uses pure MFS-like solution tools.

2.2 Band-Limited Approximation: A Two-Level Technique

Consider another circle $\Gamma_{R+2\delta}$ centered at the origin with radius $(R+2\delta)$, and seek the approximate solution of (5)–(6) as a sum of two single-layer potentials

$$u := A^{(N)}w^{(N)} + A^{(N/2)}w^{(N/2)}, \tag{25}$$

where

$$(A^{(N)}w^{(N)})(x) := \int_{\Gamma_{R+\delta}} (\log ||x-y||)\cdot w^{(N)}(y)\, d\Gamma_y$$

and

$$(A^{(N/2)}w^{(N/2)})(x) := \int_{\Gamma_{R+2\delta}} (\log ||x-y||)\cdot w^{(N/2)}(y)\, d\Gamma_y.$$

In polar coordinates, the density functions $w^{(N)}$ and $w^{(N/2)}$ are expressed in terms of finite Fourier series

$$w^{(N)}(t) = \sum_{0\neq|k|\leq N/2} \tilde{w}_k e^{ikt}, \qquad w^{(N/2)}(t) = \sum_{0\neq|k|\leq N/4} \tilde{\tilde{w}}_k e^{ikt}.$$

The Fourier coefficients are defined in such a way that the Dirichlet boundary condition (6) is satisfied i.e.

$$A^{(N)}w^{(N)} + A^{(N/2)}w^{(N/2)} = u_0. \tag{26}$$

This results in the following equation (in the sense of least squares, cf (13)):

$$u|_{\Gamma_R} = u_0 = \sum_{k\neq 0} \beta_k \cdot e^{ikt} =$$

$$= -\sum_{0\neq|k|\leq N/4} \frac{(R+\delta)\pi}{|k|}\cdot\tilde{w}_k\left(\frac{R}{R+\delta}\right)^{|k|}\cdot e^{ikt}$$

$$-\sum_{0\neq|k|\leq N/4} \frac{(R+2\delta)\pi}{|k|}\cdot\tilde{\tilde{w}}_k\left(\frac{R}{R+2\delta}\right)^{|k|}\cdot e^{ikt}$$

$$-\sum_{N/4<|k|\leq N/2} \frac{(R+\delta)\pi}{|k|}\cdot\tilde{w}_k\left(\frac{R}{R+\delta}\right)^{|k|}\cdot e^{ikt}.$$

Equation (26) has several solutions. The simplest solution technique is to solve the 'coarse level problem'

$$A^{(N/2)} w^{(N/2)} = u_0 \tag{27}$$

exactly (in the sense of least squares), yielding

$$\tilde{\tilde{w}}_k = \frac{|k|}{(R+2\delta)\pi} \cdot \left(\frac{R+2\delta}{R}\right)^{|k|} \cdot \beta_k, \qquad (0 \neq |k| \leq \frac{N}{4}). \tag{28}$$

As a next step, consider the 'fine level problem'

$$A^{(N)} w^{(N)} = u_0 - A^{(N/2)} w^{(N/2)}. \tag{29}$$

From (26) and (28), it follows that

$$\tilde{w}_k = 0 \qquad (0 \neq |k| \leq \frac{N}{4}),$$

which is interpreted as follows. The low-frequency components of the approximate solution (i.e. for which $0 \neq |k| \leq \frac{N}{4}$) are calculated from the coarse level problem (27). The high-frequency components (i.e. for which $\frac{N}{4} < |k| \leq \frac{N}{2}$) are calculated by solving (29), applying a simple (conjugate) gradient method. Since the operator $A^{(N)}$ maps the 'high-frequency subspace' $X_{N/2}^{\perp}$ spanned by the functions $\{e^{ikt} : \frac{N}{4} < |k| \leq \frac{N}{2}\}$ into itself, the speed of the convergence is determined by the condition number of the restricted operator $A^{(N)}|_{X_{N/2}^{\perp}}$, which is much less than that of the original operator $A^{(N)}$, i.e.

$$\text{cond}(A^{(N)}|_{X_{N/2}^{\perp}}) \leq \frac{|A^{(N)}_{N/4,N/4}|}{|A_{N/2,N/2}|} = 2 \cdot \left(\frac{R+\delta}{R}\right)^{N/4}.$$

But $\frac{\delta}{R} = \frac{2\pi\delta_0}{N}$ (cf (21)), which implies

$$\text{cond}(A^{(N)}|_{X_{N/2}^{\perp}}) \leq 2 \cdot e^{\frac{1}{2}\pi\delta_0}$$

and

$$\text{cond}((A^{(N)})^* A^{(N)}|_{X_{N/2}^{\perp}}) \leq 4 \cdot e^{\pi\delta_0}$$

independently of N. This means that the speed of convergence of the (conjugate) gradient method increases by a significant amount and the iteration is robust, i.e. the speed of convergence does not depend on N. In other words, the (conjugate) gradient iteration reduces the high-frequency error components quickly, thus, it can be used as an efficient *smoothing procedure* in a multi-level context.

We have obtained the following two-level algorithm:

- *Step 1*: Perform a band-limited discretization both on $\Gamma_{R+\delta}$ and on $\Gamma_{R+2\delta}$, and build up the discretized problem (26).
- *Step 2*: Solve the coarse level problem

$$A^{(N/2)} w^{(N/2)} = u_0$$

exactly (in the sense of least squares).
- *Step 3*: Apply several (conjugate) gradient iteration steps to the fine level problem

$$A^{(N)} w^{(N)} = u_0 - A^{(N/2)} w^{(N/2)}$$

in the sense of least squares, i.e. to the Gaussian normal equation of the fine level problem

$$(A^{(N)})^* A^{(N)} w^{(N)} = (A^{(N)})^* (u_0 - A^{(N/2)} w^{(N/2)}).$$

The number of the iteration steps is typically between 10 and 20, and independent of N.
- *Step 4*: If necessary, improve the approximation: redefine the coarse level problem by

$$A^{(N/2)} w^{(N/2)} = u_0 - A^{(N)} w^{(N)},$$

and continue the algorithm from Step 2.

The method requires $O(N^2)$ algebraic operations at the fine level, which is much less than the computational cost of a traditional direct solver. It should be also pointed out that no ill-conditioned problem has to be handled.

In practice, the coarse level problem needs not necessarily be solved exactly. In this case, the gradient steps may be somewhat slower.

2.3 Band-Limited Approximation: Extension to Multi-Level Method

The two-level algorithm can be extended to a multi-level technique in a straightforward way by applying the same technique to solve the coarse level problem (27).

Let us define additional concentric circles $\Gamma_{R+4\delta}$, $\Gamma_{R+8\delta}$, ..., $\Gamma_{R+2^L\delta}$. Along each circle, consider the single layer potential

$$A^{(N\cdot 2^{1-\ell})} w^{(N\cdot 2^{1-\ell})} := \int_{\Gamma_{R+2^\ell \delta}} (\log ||x-y||) \cdot w^{(N\cdot 2^{1-\ell})}(y)\, d\Gamma_y \tag{30}$$

($(\ell = 1, 2, \ldots, L)$), where, in polar coordinates, $w^{(N \cdot 2^{1-\ell})}$ has the band-limited Fourier series expansion

$$w^{(N \cdot 2^{1-\ell})}(t) := \sum_{0 \neq |k| \leq N \cdot 2^{-\ell}} \hat{w}_k^{(\ell)} \cdot e^{ikt}. \tag{31}$$

The approximate solution of (5)–(6) is sought as a two-level approximation (25)

$$u := A^{(N)} w^{(N)} + A^{(N/2)} w^{(N/2)},$$

where the density functions $w^{(N)}$ and $w^{(N/2)}$ are recursively defined as follows, using a MATLAB-style pseudocode (see also the classical multigrid techniques detailed in e.g. [17]):

$function\ [w^{(N)}] = MGC(N, A^{(N)}, w^{(N)}, b)$
$if\ \ N = N_{min}$
  $w^{(N)} := (A^{(N)})^{-1} b$
  $return$
end
$w^{(N/2)} := 0$
$for\ \ i = 1 : iteration\_number$
  $w^{(N/2)} := MGC(N/2, A^{(N/2)}, w^{(N/2)}, u_0 - A^{(N)} w^{(N)})$
  $w^{(N)} := gradient\_steps(N, A^{(N)}, w^{(N)}, u_0 - A^{(N/2)} w^{(N/2)})$
end
$return$

In practice, the use of the band-limited approximation is inconvenient, especially when the shape of the domain is more complicated. However, after minor modifications, the idea still can be applied to the case of usual point sources as shown in the next subsection.

2.4 Approximation by Near-Boundary Point Sources

Let $N \in \mathbf{N}$ be a fixed even number (a power of 2 in the following). Now let us look for the approximate solution of (5)–(6) in terms of point sources located equidistantly along $\Gamma_{R+\delta}$, i.e.

$$u^{(N)}(x) := \sum_{j=0}^{N-1} \alpha_j \cdot \log ||x - x_j||, \tag{32}$$

where x_j is the jth source point

$$x_j = (R + \delta) \cdot \left(\cos \frac{2j\pi}{N}, \sin \frac{2j\pi}{N} \right) \qquad (j = 0, 1, \ldots, N-1).$$

We assume again that the sources are located in the vicinity of Γ_R, i.e. the condition (21) is still satisfied. Thus, $\frac{\delta}{R} = \frac{2\pi}{N} \cdot \delta_0$ for some $\delta_0 > 0$ which is independent of N.

Standard calculations show that the function $u^{(N)}$ can be expressed in terms of Fourier series in the following way (written in polar coordinates). We have

$$u^{(N)}(r,t) = (\log(R+\delta)) \cdot \left(\sum_{j=0}^{N-1} \alpha_j \right) - \frac{1}{2} \cdot \sum_{k \neq 0} \frac{1}{|k|} \left(\frac{r}{R+\delta} \right)^{|k|} \hat{\alpha}_k \cdot e^{-ikt}, \qquad (33)$$

where $\hat{\alpha}_k$ $(k = 0, 1, \ldots, N-1)$ denotes the discrete Fourier transform of the finite sequence of the coefficients α_j $(j = 0, 1, \ldots, N-1)$

$$\hat{\alpha}_k := \sum_{j=0}^{N-1} \alpha_j e^{\frac{2\pi i k j}{N}} \qquad (k = 0, 1, \ldots N-1).$$

Here, $\hat{\alpha}_k$ is considered to be extended to the set $\mathbf{Z}$ of all integers in an N-periodic way.

Thus, the approximate solution along the boundary Γ_R can be expressed as

$$u^{(N)}|_{\Gamma_R} = u^{(N)}(R,t) = \hat{\alpha}_0 \cdot \log(R+\delta) - \frac{1}{2} \cdot \sum_{k \neq 0} \frac{1}{|k|} \left(\frac{R}{R+\delta} \right)^{|k|} \hat{\alpha}_k \cdot e^{-ikt}. \qquad (34)$$

Define the discrete operator $A^{(N)} : \mathbf{C}^N \to L_2(0, 2\pi)$ by

$$(A^{(N)} \boldsymbol{\alpha})(t) := u^{(N)}(R,t),$$

where $\boldsymbol{\alpha} \in \mathbf{C}^N$ denotes the vector formed by the components α_j $(j = 0, 1, \ldots, N-1)$. For determining the approximate solution $u^{(N)}$, one has to solve the system of equations

$$A^{(N)} \boldsymbol{\alpha} = u_0 \qquad (35)$$

in the sense of least squares, i.e. the Gaussian normal equations

$$(A^{(N)})^* A^{(N)} \boldsymbol{\alpha} = (A^{(N)})^* u_0. \qquad (36)$$

Now we will show that the matrix $(A^{(N)})^* A^{(N)}$ is a diagonal matrix in the orthonormal basis formed by the vectors $\boldsymbol{\alpha}^{(-\frac{N}{2}+1)}, \boldsymbol{\alpha}^{(-\frac{N}{2}+2)}, \ldots, \boldsymbol{\alpha}^{(\frac{N}{2})} \in \mathbf{C}^N$, where

$$\boldsymbol{\alpha}_j^{(p)} := \frac{1}{\sqrt{N}} \cdot e^{-\frac{2\pi i p j}{N}}$$

$(j = 0, 1, \ldots, N-1)$, $(p = -\frac{N}{2}+1, -\frac{N}{2}+2, \ldots, \frac{N}{2})$. By definition, it is easy to see that

$$(\hat{\boldsymbol{\alpha}}^{(p)})_k = \begin{cases} \sqrt{N} & \text{if } k \equiv p \pmod N, \\ 0 & \text{otherwise,} \end{cases}$$

which implies that, for $p \neq 0$,

$$A^{(N)}\boldsymbol{\alpha}^{(p)} = (\hat{\alpha}^{(p)})_0 \cdot \log(R+\delta) - \frac{1}{2} \cdot \sum_{k \neq 0} \frac{1}{|k|} \left(\frac{R}{R+\delta}\right)^{|k|} (\hat{\boldsymbol{\alpha}}^{(p)})_k \cdot e^{-ikt} =$$

$$= -\frac{\sqrt{N}}{2} \cdot \sum_{\ell=-\infty}^{+\infty} \frac{1}{|p+\ell N|} \left(\frac{R}{R+\delta}\right)^{|p+\ell N|} \cdot e^{-i(p+\ell N)t}.$$

If $p = 0$, then

$$A^{(N)}\boldsymbol{\alpha}^{(0)} = (\hat{\alpha}^{(0)})_0 \cdot \log(R+\delta) - \frac{1}{2} \cdot \sum_{k \neq 0} \frac{1}{|k|} \left(\frac{R}{R+\delta}\right)^{|k|} (\hat{\boldsymbol{\alpha}}^{(0)})_k \cdot e^{-ikt} =$$

$$= \sqrt{N} \cdot \log(R+\delta) - \frac{\sqrt{N}}{2} \cdot \sum_{\ell \neq 0} \frac{1}{|\ell N|} \left(\frac{R}{R+\delta}\right)^{|\ell N|} \cdot e^{-i\ell N t}.$$

Since the (p, q)th element of the matrix $(A^{(N)})^* A^{(N)}$ is

$$((A^{(N)})^* A^{(N)})_{pq} = \langle (A^{(N)})^* A^{(N)} \boldsymbol{\alpha}^{(p)}, \boldsymbol{\alpha}^{(q)} \rangle_{\mathbf{C}^N} = \langle A^{(N)} \boldsymbol{\alpha}^{(p)}, A^{(N)} \boldsymbol{\alpha}^{(q)} \rangle_{L_2(0,2\pi)},$$

the above equalities imply that the matrix $(A^{(N)})^* A^{(N)}$ is a diagonal matrix, as stated above. Moreover, the diagonal entries can easily be calculated by virtue of Parseval's theorem. For $k \neq 0$

$$((A^{(N)})^* A^{(N)})_{kk} = 2N\pi \cdot \sum_{\ell=-\infty}^{+\infty} \frac{1}{4|k+\ell N|^2} \left(\frac{R}{R+\delta}\right)^{2|k+\ell N|}$$

$$=: 2N\pi \cdot \sum_{\ell=-\infty}^{+\infty} c_{k+\ell N}^2,$$

where

$$c_\gamma^2 := \frac{1}{4|\gamma|^2} \left(\frac{R}{R+\delta}\right)^{2|\gamma|}.$$

For $k = 0$

$$((A^{(N)})^* A^{(N)})_{00} = 2N\pi \cdot \left((\log(R+\delta))^2 + \sum_{\ell \neq 0} \frac{1}{4|\ell N|^2} \left(\frac{R}{R+\delta} \right)^{2|\ell N|} \right).$$

We have obtained that for nonzero indices k (written in a more convenient form)

$$\begin{aligned} &((A^{(N)})^* A^{(N)})_{kk} = \\ &= 2N\pi \cdot (c_k^2 + (c_{k+N}^2 + c_{k+2N}^2 + \ldots) + (c_{k-N}^2 + c_{k-2N}^2 + \ldots)). \end{aligned} \tag{37}$$

From (37), the condition number can be conveniently estimated both in the whole space and also in the 'high frequency subspace', similarly than in the case of the band-limited approximation.

2.4.1 Condition Number

The diagonal entries of the matrix $(A^{(N)})^* A^{(N)}$ should be estimated from below and above. Obviously

$$((A^{(N)})^* A^{(N)})_{kk} \geq 2N\pi \cdot c_k^2 \geq 2N\pi \cdot \frac{1}{N^2} \cdot \left(\frac{R}{R+\delta} \right)^{2 \cdot \frac{N}{2}},$$

and therefore

$$((A^{(N)})^* A^{(N)})_{kk} \geq \frac{1}{N^2} \cdot 2N\pi \cdot e^{-2\pi\delta_0} \tag{38}$$

for $k \neq 0$. On the other hand

$$((A^{(N)})^* A^{(N)})_{kk} \leq 2N\pi \cdot (c_k^2 + (c_{N/2}^2 + c_{3N/2}^2 + \ldots) + (c_{-N/2}^2 + c_{-3N/2}^2 + \ldots))$$

$$\leq 2N\pi \left[\tfrac{1}{4k^2} \left(\tfrac{R}{R+\delta} \right)^{2|k|} + \tfrac{2}{N^2} \left(\tfrac{R}{R+\delta} \right)^{N} \cdot \left(\tfrac{1}{1^2} + \tfrac{1}{3^2} + \tfrac{1}{5^2} + \ldots \right) \right],$$

whence

$$((A^{(N)})^* A^{(N)})_{kk} \leq 2N\pi \cdot \left(\frac{1}{4} + \frac{\pi^2}{4N^2} \right). \tag{39}$$

Thus, the condition number can be estimated as

$$\mathrm{cond}((A^{(N)})^* A^{(N)}) \leq N^2 \cdot \frac{\pi^2 + 1}{4} \cdot e^{2\pi\delta_0} \tag{40}$$

similarly to (24) in the band-limited approximation, and this estimation does not change essentially when the 0th diagonal entry is also taken into account. This means that the condition number increases moderately with N, but makes the speed of convergence of the gradient method low. However, similarly to the band-limited case, the high-frequency error may decrease much faster. To show this, it is sufficient to estimate the condition number of the matrix $((A^{(N)})^* A^{(N)})$ restricted to the 'high-frequency subspace' i.e. for the indices k for which

$$\frac{N}{4} < |k| \le \frac{N}{2}$$

is satisfied. The lower bound of the diagonal entries is the same as previously, i.e.

$$((A^{(N)})^* A^{(N)})_{kk} \ge 2N\pi \cdot \frac{1}{N^2} \cdot \left(\frac{R}{R+\delta}\right)^N,$$

while the estimation of the upper bound is slightly different, i.e.

$$((A^{(N)})^* A^{(N)})_{kk} \le 2N\pi \cdot (c_{N/4}^2 + (c_{N/2}^2 + c_{3N/2}^2 + \ldots) + (c_{-N/2}^2 + c_{-3N/2}^2 + \ldots))$$

$$\le 2N\pi \left[\frac{4}{N^2} \left(\frac{R}{R+\delta}\right)^{2\cdot\frac{N}{4}} + \frac{\pi^2}{4N^2} \left(\frac{R}{R+\delta}\right)^N \right],$$

which implies that in the 'high-frequency subspace'

$$\mathrm{cond}((A^{(N)})^* A^{(N)}) \le \frac{4\left(\frac{R}{R+\delta}\right)^{N/2} + \frac{\pi^2}{4}\left(\frac{R}{R+\delta}\right)^N}{\left(\frac{R}{R+\delta}\right)^N} \le 4 \cdot e^{\pi\delta_0} + \frac{\pi^2}{4} \tag{41}$$

independently of N. That is, the (conjugate) gradient method is a suitable smoothing procedure in the multi-level context, similarly to the band-limited case, thus, the two-level as well as the multi-level method of the preceding subsections can be built up without difficulty, using the discretization of (35) instead of (17). For illustration, Fig. 1 shows the arrangements of the sources both on the fine level ($x_j^{(1)}$, $j = 0, 1, \ldots, N-1$) and on the coarse level as well ($x_j^{(2)}$, $j = 0, 1, \ldots, \frac{N}{2} - 1$). Note that this type of approximation can be implemented much easier than the band-limited technique and can be generalized to more general domains in a natural way.

2.5 *Definition of Sources Using Quadtrees*

As seen in the previous subsections, the 'coarse problem' requires half as much source points as the 'fine problem', located at double distance from the boundary.

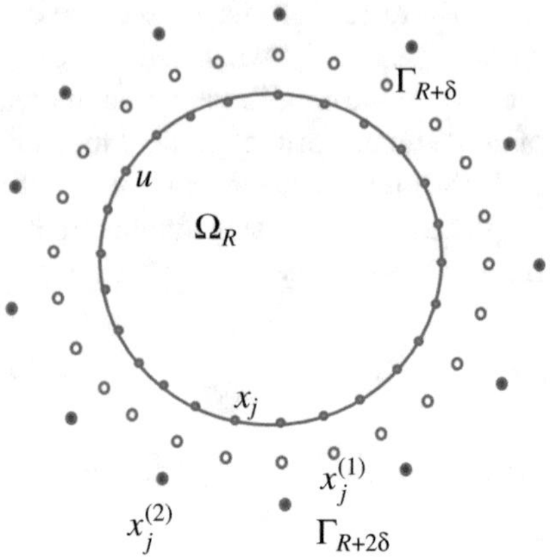

Fig. 1 Two-level MFS, locations of sources

In the multi-level expansion, this means that the density of the spatial distribution of the source points should decrease exponentially when the distance from the boundary increases. This can be performed by the quadtree (octtree in 3D) algorithm in a completely automatic way. Recall that the quadtree algorithm is a recursively defined, systematic subdivision of an initial square into four congruent subsquares controlled by a finite number of points (referred to as *controlling points* hereafter). A subdivision is performed if the number of controlling points contained in the actual subsquare exceeds a predefined minimal value, provided that the level of subdivision remains under a predefined maximal value. By performing some additional subdivisions, it can be assured that the ration of the neighbouring cell sizes is at most 2 (regularization of the quadtree cell systems), i.e. no abrupt changes in neighbouring cell sizes occur. This procedure results in a non-equidistant, non-uniform cell system which automatically exhibits local refinements in the vicinity of the controlling points. This cell system is suitable for defining simple finite volume schemes as well (see [6]). In the presented multi-level technique, the quadtree cell system is controlled by the boundary collocation points $x_1, x_2, \ldots, x_M \in \Gamma_R$.

As an example, Fig. 2 shows a regular quadtree cell system controlled by the boundary points of a circle centered at the point (0.5, 0.5) with radius $R = 0.3$. Figure 3 shows all the external sources. In Fig. 4, the sources belonging to the different levels are displayed. The maximal subdivision level is 8, i.e. the size of the finest cell is $1/256$. The level of the finest (coarsest) source points is 7 (3, respectively). The number of boundary collocation points is 476; the collocation points are the centers of the cells at the 8th i.e. the finest subdivision level, which contain some boundary points.

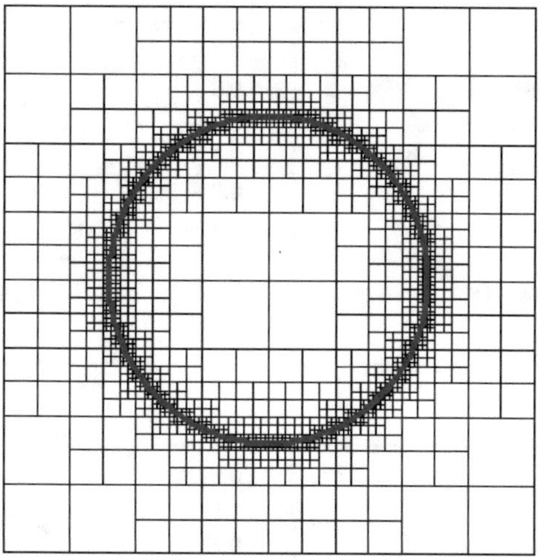

Fig. 2 Quadtree cell system generated by a circle

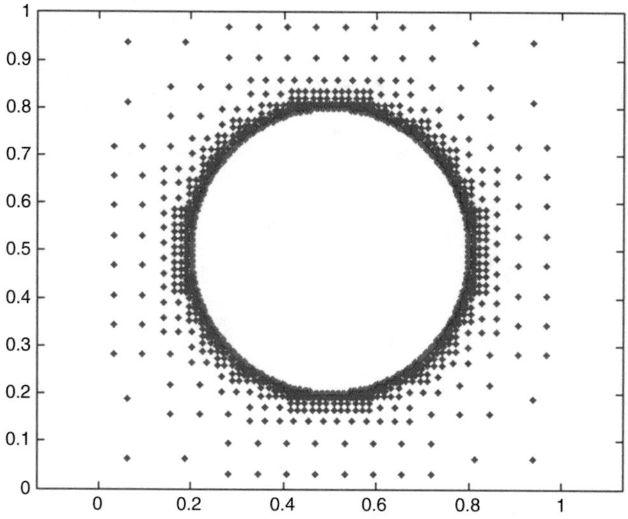

Fig. 3 Circle, the source point locations

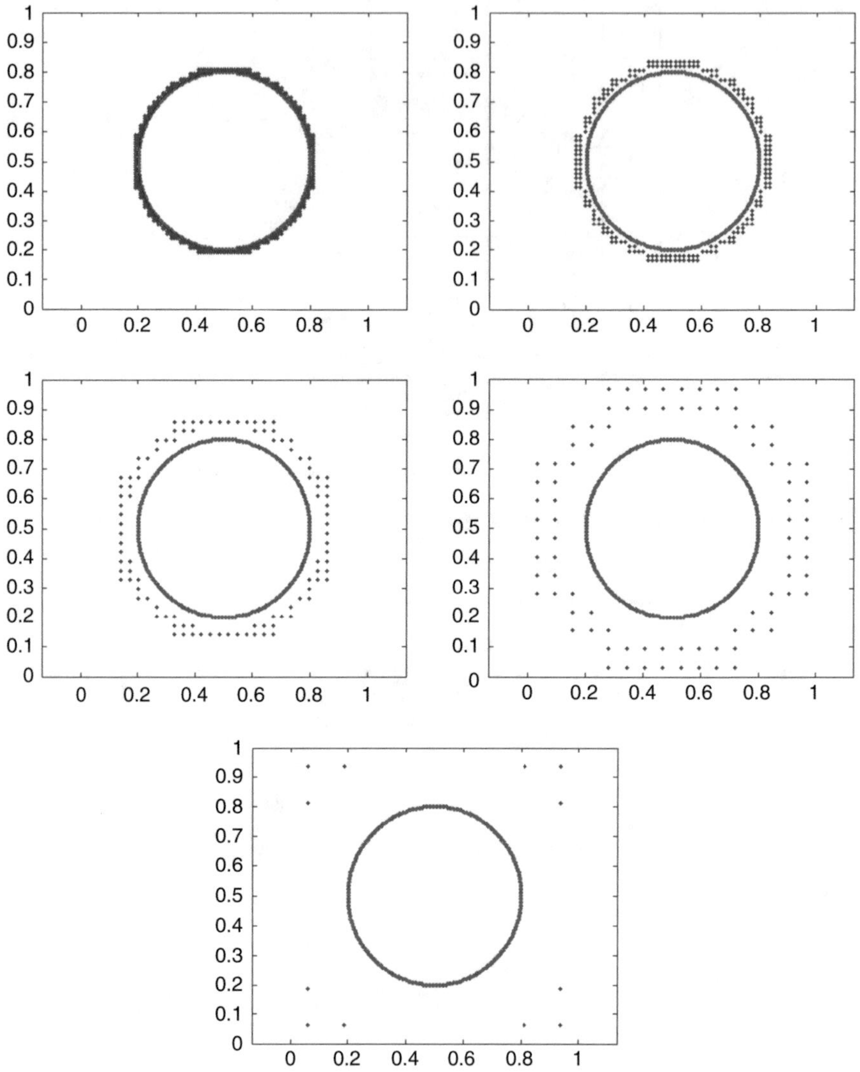

Fig. 4 Circle, the source point locations at the different levels

3 A Numerical Example

The above presented technique is illustrated through a numerical example. Consider the problem (5)–(6) in the unit circle Ω_1 with the test solution

$$u(x, y) = (\cos 2\pi x) \cdot (\sinh 2\pi y), \quad (42)$$

where the Dirichlet boundary condition is defined to be consistent with the test solution (42). Discretize the boundary Γ_1 by N equidistantly spaced boundary collocation points

$$(x_j, y_j) := \left(\cos\frac{2j\pi}{N}, \sin\frac{2j\pi}{N}\right) \qquad (j = 0, 1, \ldots, N-1).$$

Let $\Gamma_{1+\delta}$ be a circle centered at the origin with radius $(1+\delta)$, where $\delta := \frac{2\pi}{N}$. Define the source points as

$$(\tilde{x}_j, \tilde{y}_j) := (1+\delta)\cdot\left(\cos\frac{2j\pi}{N}, \sin\frac{2j\pi}{N}\right) \qquad (j = 0, 1, \ldots, N-1),$$

i.e. they are 'near-boundary sources' in the sense of (21). First, we have computed the standard MFS-solutions with different values of N, while the number of boundary collocation points was unchanged, $M = 256$. The solutions of the corresponding Gaussian normal equations were calculated by a standard direct method. Table 1 shows the relative L_2-norms of the residuals $(A^{(N)}\alpha_N - u_0)$, where α_N is the solution of (35) (in the sense of least squares) and $A^{(N)}$ is the discrete MFS-operator defined by (32). Table 1 also contains the condition numbers of $A^{(N)}$ and of $(A^{(N)})^* A^{(N)}$ as well.

Table 1 illustrates that the condition numbers increase moderately with N, however, this makes the familiar iterative methods slow. That is, the use of the standard direct methods is needed, but the number of necessary algebraic operations is rather high, $O(N^3)$.

A two-level method was also applied to the above test problem by using one coarse level, as illustrated in Fig. 1. The coarse level problem was solved directly, while, to the fine level problem, 16 conjugate gradient iterations were applied, and the fine-coarse level improvement steps were repeated iteratively. The errors were always calculated on the fine level. The relative L_2-norms of the residuals for different values of N can be seen in Table 2. Here the coarse level problem contains N sources, while the fine level possesses $2N$ sources. Table 3 shows the relative L_2-norms of the residuals, when, for the coarse level problem, 16 conjugate gradient iteration steps were applied instead of solving the coarse problem directly.

Comparing these results with those of Table 2, one can observe that the accuracy is more or less the same as in the previous case. The situation remains similar, when

Table 1 Relative L_2-norms of the residuals and the condition numbers

N	16	32	64	128	256
cond$(A^{(N)})$	57.4	166.5	412.5	925.7	1390.8
cond$((A^{(N)})^* A^{(N)})$	3304	27730	170176	856981	1934341
Relative L_2-error on Γ_1 (%)	11.68	1.775	0.1203	0.0250	0.0148

Test solution: (42), method: single-level MFS

Table 2 Relative L_2-norms of the residuals, two-level MFS

$N/2N$	8/16	16/32	32/64	64/128	128/256
Relative L_2-error (%)	12.01	1.065	0.1351	0.0200	2.1e−8

Test solution: (42), method: two-level MFS. Direct solution on the coarse level

Table 3 Relative L_2-norms of the residuals, two-level MFS

$N/2N$	8/16	16/32	32/64	64/128	128/256
Relative L_2-error (%)	12.01	1.065	0.1351	0.0202	0.0007

Test solution: (42), method: two-level MFS. Conjugate gradient iterations on the coarse level

Table 4 Relative L_2-norms of the residuals, three-level MFS

$N/2N/4N$	4/8/16	8/16/32	16/32/64	32/64/128	64/128/256
Relative L_2-error (%)	12.01	1.106	0.0541	0.0058	4.3e−5

Test solution: (42), method: three-level MFS

Table 5 Relative L_2-norms of the residuals, single-level MFS on quadtree cell system

L	3	4	5	6	7
N	12	88	104	216	376
Relative L_2-error (%)	58.58	0.0086	0.0011	0.0006	0.0048

Test solution: (43), method: single-level MFS on quadtree cell system

applying a multi-level technique using several consecutively coarser levels. Table 4 shows 3-grid results (with N coarsest and $4N$ finest source points). Note also that the presented technique does not suffer from the problem of severely ill-conditioned matrices.

Finally, some illustration of the automatic quadtree-based source definition technique is shown. Now let the domain Ω be a circle contained in the unit square centered at the point (0.5, 0.5) with radius 0.3 as shown in Figs. 2 and 3. The number of boundary collocation points is 476 in each level (the centers of the finest cells which contain some boundary points). The source points belonging to the different levels can be seen in Fig. 4. Their quadtree subdivision levels vary from 3 to 7, and the numbers of the corresponding points are 12, 88, 104, 216 and 376. Now the test solution is defined by

$$u(x, y) = \cos \frac{2\pi(x-0.5)}{0.3} \cdot \sinh \frac{2\pi(y-0.5)}{0.3}. \tag{43}$$

Table 5 shows the relative L_2-norms of the residuals (calculated in the boundary collocation points). Here L denotes the quadtree subdivision level and N is the number of source points in the corresponding level. The approximate solutions were calculated by a standard direct method. This indicates that the accuracy remains good enough, but the systems are severely ill-conditioned. From level 4, the condition numbers are above 10^{12}. However, using a two-level technique (applying 20 conjugate gradient steps in both levels), we obtain an acceptable accuracy,

Table 6 Relative L_2-norms of the residuals, two-level MFS on quadtree cell system

L_1/L_2	3/4	4/5	5/6	6/7
N_1/N_2	12/88	88/104	104/216	216/376
Relative L_2-error (%)	0.2729	0.0601	0.0263	0.0183

Test solution: (43), method: two-level MFS on quadtree cell system

with much less computational cost and without having to deal with severely ill-conditioned matrices, as shown in Table 6. Here L_1 denotes the quadtree level of the coarser problem, L_2 is that of the finer problem. N_1 (resp. N_2) is the number of sources in the coarse (resp. the fine) level. The norms of the residuals i.e. the errors are calculated always in the fine level.

4 Summary and Conclusions

A traditional form of the Method of Fundamental Solutions was applied. The external source point locations were defined in a completely automated way using a quadtree/octtree subdivision technique controlled by the boundary collocation points. Thus, the density of the spatial distribution of the sources decreases far from the boundary. The external sources form groups of points which are interpreted as coarse grids and fine grids (however, they have no grid structure at all). At each level, the approximate solution was defined by least square approximation of the corresponding Gaussian normal equations. As a smoothing procedure, a simple (conjugate) gradient method was used. It was shown that, using the above groups of source points, the gradient method significantly reduces the high-frequency components of the errors, which made it possible to build up a multi-level method in a simple way. The resulting method has acceptable accuracy and much less computational complexity compared to the usual solution techniques. No special tricks to treat singularities are needed. At the same time, the problem of solving large and severely ill-conditioned linear systems is completely avoided.

Acknowledgements The research was partly supported by the European Union in the framework of the project GINOP-2.3.4-15-2016-00003.

References

1. C.J.S. Alves, C.S. Chen, B. Šarler, The method of fundamental solutions for solving Poisson problems, in *Proceedings of the 24th International Conference on the Boundary Element Method incorporating Meshless Solution Seminar*. International Series on Advances in Boundary Elements, vol. 13, ed. by C.A. Brebbia, A. Tadeu, V. Popov (WitPress, Southampton, 2002), pp. 67–76

2. W. Chen, F.Z. Wang, A method of fundamental solutions without fictitious boundary. Eng. Anal. Bound. Elem. **34**, 530–532 (2010)
3. W. Chen, L.J. Shen, Z.J. Shen, G.W. Yuan, Boundary knot method for Poisson equations. Eng. Anal. Bound. Elem. **29**, 756–760 (2005)
4. C.S. Chen, A. Karageorghis, Y. Li, On choosing the location of the sources in the MFS. Numer. Algorithms **72**, 107–130 (2016)
5. G.S.A. Fam,Y.F. Rashed, A study on the source points locations in the method of fundamental solutions, in *Proceedings of the 24th International Conference on the Boundary Element Method incorporating Meshless Solution Seminar* (17–19 June 2002 Sintra, Portugal). International Series on Advances in Boundary Elements, vol. 13, ed. by C.A. Brebbia, A. Tadeu, V. Popov (WitPress, Southampton, 2002), pp. 297–312
6. C. Gáspár, A meshless polyharmonic-type boundary interpolation method for solving boundary integral equations. Eng. Anal. Bound. Elem. **28**, 1207–1216 (2004)
7. C. Gáspár, Some variants of the method of fundamental solutions: regularization using radial and nearly radial basis functions. Cent. Eur. J. Math. **11/8**, 1429–1440 (2013)
8. C. Gáspár, A regularized multi-level technique for solving potential problems by the method of fundamental solutions. Eng. Anal. Bound. Elem. **57**, 66–71 (2015)
9. C. Gáspár, A multi-level meshless method based on an implicit use of the method of fundamental solutions, in Proceedings of the 6th International Conference on Computational Methods, 14th - 17th July, 2015, Auckland (ScienTech Publisher, Mason, 2015), Paper ID: 1032 (ISSN 2374–3948 online)
10. M.A. Golberg, The method of fundamental solutions for Poisson's equation. Eng. Anal. Bound. Elem. **16**, 205–213 (1995)
11. Y. Gu, W. Chen, J. Zhang, Investigation on near-boundary solutions by singular boundary method. Eng. Anal. Bound. Elem. **36**, 1173–1182 (2012)
12. X. Li, On convergence of the method of fundamental solutions for solving the Dirichlet problem of Poisson's equation. Adv. Comput. Math. **23**, 265–277 (2005)
13. Y. Saad, *Iterative Methods for Sparse Linear Systems: Second Edition*. Applied Mathematics, vol. 82 (SIAM, Chicago, 2003)
14. B. Šarler, A modified method of fundamental solutions for potential flow problems, in *The Method of Fundamental Solutions - A Meshless Method*, ed. by C.S. Chen, A. Karageorghis, Y.S. Smyrlis (Dynamic Publishers, Inc., Atlanta, 2008), pp. 299–321
15. B. Šarler, Solution of potential flow problems by the modified method of fundamental solutions: formulations with the single layer and the double layer fundamental solutions. Eng. Anal. Bound. Elem. **33**, 1374–1382 (2009)
16. R. Schaback, Adaptive numerical solution of MFS systems, in *The Method of Fundamental Solutions - A Meshless Method*, ed. by C.S. Chen, A. Karageorghis, Y.S. Smyrlis (Dynamic Publishers, Inc., Atlanta, 2008), pp. 1–27
17. K. Stüben, U. Trottenberg, Multigrid methods: fundamental algorithms, model problem analysis and applications, in GDM-Studien, vol. 96, Birlinghoven (1984)
18. D.L. Young, K.H. Chen, C.W. Lee, Novel meshless method for solving the potential problems with arbitrary domain. J. Comput. Phys. **209**, 290–321 (2005)

Explicit Margin of Safety Assessment of Composite Structure

J. H. Gosse and E. J. Sharp

Abstract In this paper, we discuss the assessment of the margins of safety (MOS) for engineered structures (buildings, bridges, machines and aircraft). With respect to metallic structures the MOS are primarily a function of the observed yield stress of the metal (critical property of the constitutive material). With continuous fiber reinforced polymer composites (composites in this paper) such a crisp measurement of the "yield stress" is not available. The result has been an empirical test-intensive building block approach to the assessment of the MOS for composite structure. However, the capability to use the critical properties of the constitutive materials of the composite system to evaluate the MOS of the composite structure is now possible (explicit MOS assessment). Such an approach will lead to significant reductions in cost and time-to-design as well as a practical means towards expanding the design space. The measurement of the critical measures, de-homogenization of homogenous strain states and classical convergence of the numerical solutions involved are discussed in detail. Finally, integrating generalized finite elements into the analysis process will allow for rapid and efficient assessment of the MOS for global structures.

1 Introduction

All engineered structures need to be safe. The MOS to assess safety is primarily a function of the peak value of the measure of interest (for a given load case) at a point within the structure and the critical material property of the material used in the structure. For example, the MOS can be expressed very simply [1],

$$\text{MOS} = \frac{\sigma_{\text{tu}}}{\sigma_t} - 1 \tag{1}$$

J. H. Gosse (✉) · E. J. Sharp
MSC Software, Bellevue, WA, USA
e-mail: jonathan.gosse@mscsoftware.com; joe.sharp@mscsoftware.com

M. Griebel, M. A. Schweitzer (eds.), *Meshfree Methods for Partial Differential Equations IX*, Lecture Notes in Computational Science and Engineering 129,
https://doi.org/10.1007/978-3-030-15119-5_3

where σ_{tu} is the allowable stress and σ_t is the in-situ stress. Actual expressions of the MOS for bridges and buildings [2] as well as machinery [3] and aircraft [1] are usually more complicated than that shown in (1). MOS that pertain to stability or crippling checks usually do not involve the critical measures of the material.

All solid materials (systems that resist shear) deform either by dilatation, distortion or both. The mechanisms do not synergize (they are not additive at a point, they compete). Mathematically, these two deformations can be expressed as [4],

$$\varepsilon_{\text{dilatation}} = \frac{\nabla V}{V} = (1 + \varepsilon_1)(1 + \varepsilon_2)(1 + \varepsilon_3) - 1 \tag{2}$$

which is the expression describing change in volume (as a function of the principal strains). For small strains (2) simplifies to,

$$\varepsilon_{\text{dilatation}} \cong \varepsilon_1 + \varepsilon_2 + \varepsilon_3 \tag{3}$$

The other deformation mode is distortion (sometimes called the deviatoric strain) [4],

$$\varepsilon_{\text{distortion}} = \sqrt{\frac{1}{6}\left[(\varepsilon_1 - \varepsilon_1)^2 + (\varepsilon_1 - \varepsilon_3)^2 + (\varepsilon_2 - \varepsilon_3)^2\right]} \tag{4}$$

which is the square root of the second invariant of the strain deviator tensor. Note that (4) is a more fundamental form of the von Mises strain where the factor $\frac{1}{2}$ is replaced with the factor $\frac{1}{6}$. As can be seen in (1) either of these factors would cancel in the assessment of the MOS. The use of these two measures to assess material failure is the basis of the Onset theory. The von Mises yield criterion is the first published instance of the theory. The measures of interest for solids are dilatation and distortion. Their critical values are the critical material properties of the solid. All solids possess these two critical measures. Deformations are fundamentally functions of strain, therefore strain invariants, Eqs. (2)–(4) are the measures to use in the assessment of the MOS. If the elastic or plastic modulus is known precisely as a function of both space and time, then the stress invariants may be admitted as well. This is never the case with viscoelastic or plastic solids [5].

Typical stress-strain behaviors for uniaxially loaded test coupons for both a ductile metal (example, a ferrous alloy) and general laminated carbon fiber reinforced polymers (CFRP) are shown in Fig. 1. Typically, CFRP have around 60% effective fiber volume resulting in the polymer being constrained [6]. Therefore, the stress-strain behavior shown in Fig. 1 for ductile metals does not manifest itself in constrained polymers, once the critical deformation exists the polymer will cavitate. As can be seen in Fig. 1, the ductile metal is reasonably linear in the stress-strain response until the material yields. Further observations exist as well, but it is the initial yielding of the material that is of most importance with respect to the MOS. Within the composite coupon, the constituent materials are also reaching their

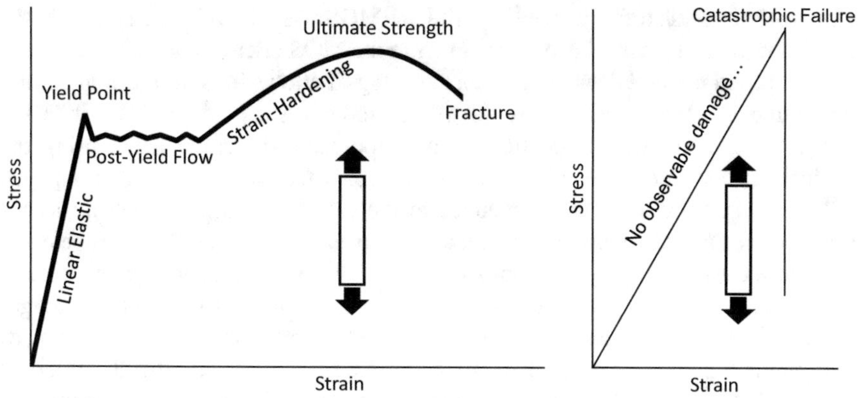

Fig. 1 Stress-strain behavior of an uniaxially loaded ductile metal coupon and the stress-strain behavior of an uniaxially loaded laminated CFRP coupon

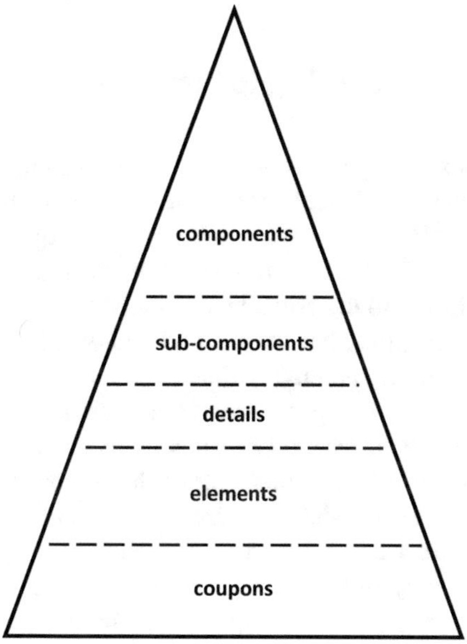

Fig. 2 Building block approach to design in composite structure

critical limits with respect to dilatation and distortion, but this is not obvious in Fig. 1. It is this final catastrophic failure mode along with an ever-changing failure load with changes in the laminate layup (stacking sequence of the plies within the laminate) that has led to the building block approach illustrated in Fig. 2.

Any change in the laminate layups, material systems, geometric configurations, loading/boundary conditions or the environment requires that the complete set of tests shown in Fig. 2 be repeated. In addition, it is the coupons that establish the allowable stresses/strains needed to write the MOS. As with metallic structure, the use of the critical material properties to write the MOS circumvents this procedure. The proposed method does not eliminate testing, it allows for smart testing greatly mitigating the test requirements relative to today's standard practice. It is the potential to return to a more fundamental approach to writing MOS using the intrinsic critical properties of the materials that is the main theme of this paper. All sources of strain are to be included in the calculation of the strain invariants (applied mechanical, thermal residual, clamp-up, environmental, shrinkage, etc.).

Before the critical property approach to writing MOS can be implemented there are issues that must be addressed first including; de-homogenization of the homogenous strain states from analytical or numerical solutions, classical convergence of the laminate solution (removal of mesh dependency from discretized numerical solutions), the use of the critical measures to write the MOS and implementation of the generalized finite element method to address global structural assessments subject to hundreds of load cases and thousands of structural details.

2 De-homogenization of the Homogenous Strain State

With respect to composite material systems the constituent material properties must be homogenized to obtain analytical or numerical solutions. In most cases, it is not practical nor even possible to explicitly model the constituents within the global structure. To write the MOS, the in-situ strain states need to be de-homogenized so that they may be compared to the critical constituent material properties. There are many ways to accomplish this, but the method addressed in this paper is that by Ritchey et al. [7]. The process is roughly illustrated in Fig. 3.

The homogenization of composite material systems primarily involves the effective elastic moduli, the effective Poisson's ratios and the effective coefficients of thermal expansion [8] (effective meaning of the ply or average). The resulting homogenous strains are then de-homogenized for use in the writing of the MOS. As shown in Fig. 3, there are two 3D unit cell models of interest, one for the least efficient packing efficiency (uniform spacing of the fibers) and one for the most efficient packing efficiency (hexagonal spacing of the fibers). These two extremes tend to capture the spatial variability within composite material systems represented in Fig. 3. Critical locations are identified within each of the unit cells for de-

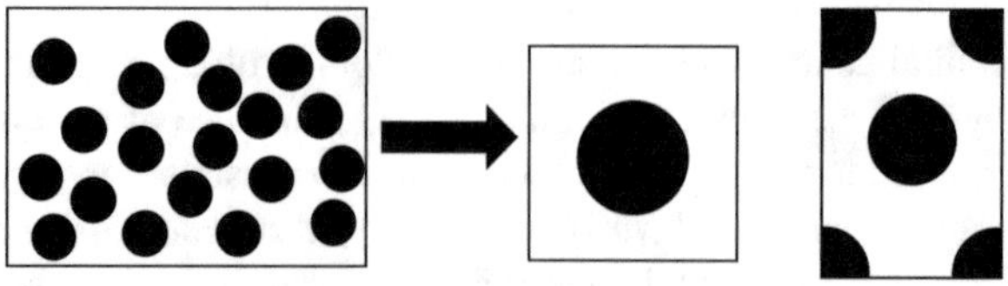

Fig. 3 De-homogenization of the composite material system

homogenization of the homogenous strain state for a given gauss point in the global structure. The resulting strain invariants of greatest magnitude are retained to write the MOS at the selected gauss point. The detailed procedure for de-homogenization is provided in [7] and involves interrogation of all constituent phases within the composite material system. The procedures provided in [7] have been investigated by The Boeing Company, The MSC Software Company, The University of New South Wales (Australia), Hanyang University (South Korea), Delft University (The Netherlands), Purdue University (U.S.) and U.S. Naval Air Systems Command.

3 Classical Convergence of the Laminate Solution (Primarily a Finite Element Solution)

Laminate systems have issues regarding convergence that homogenous systems do not. In addition to convergence issues associated with changes in geometry and applied loads with small footprints, laminate structure also has convergence issues with respect to laminate free-edges (traction-free surfaces). Measures such as distortion, stresses and strains normal to ply planes and transverse shear stresses and strains tend to be vertically asymptotic as mesh discretization is refined near the free-edge. If the trend towards vertically asymptotic behavior at the free-edge is smooth, then there is no indication from the solution sets (sets defining convergence) as to when to cease refining the mesh [9]. The two deformation modes are the most appropriate set of measures to use to establish convergence when using the Onset theory (2–4). The above argument suggests that de-homogenized dilatation should be used to establish convergence since distortion tends to be either divergent or convergent-divergent. Numerical experiments using halving [10] support this assumption. The use of halving is illustrated in Fig. 4.

Halving is illustrated in Fig. 4 using a hole within a laminate. The critical element dimension is that normal to the free-edge (in this case the free-edge is curved). Halving is established on this dimension; all other dimensions are a function of valid finite element design in mechanics. In 3D halving results in elements divided into eighths. Most of the published literature regarding laminate convergence at a free-edge involves long straight edges only. In this case, dilatation will converge absolutely, and distortion will smoothly diverge. In practice dividing all elements

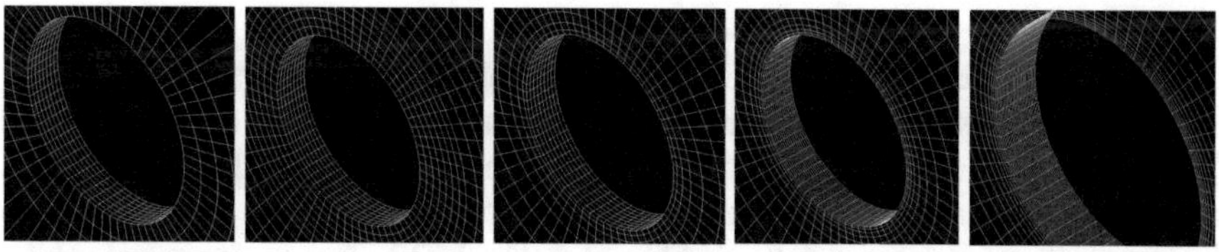

Fig. 4 Solution sets as a function of halving to establish convergence

within a numerical model into eighths is not practical (however, the concept of $h_{\max} \rightarrow 0$ is technically most accurate [10]) so local biasing is usually employed [10]. Over-meshing should be avoided otherwise convergence may be established but the limit may be incorrect (drifting) [10]. The authors have found that once the convergence process results in less than a 5% change in the de-homogenized dilatation at the free-edge the exercise can be terminated and the resulting distortion accepted as well (this last assumption must be proven with analysis and test as has been established by the authors). With curved free-edges, both measures may converge at first and then; both diverge, one diverges, and the other does not, or both converge. If both measures diverge then the measure that diverges first establishes the mesh (divergence is an indication that the mathematical solution to the partial differential is correct, but it may no longer have physical meaning (example, measures of infinite value do not exist). If both measures converge then the 5% rule from above is used. If one measure converges and the other does not, then the mesh that establishes divergence is the mesh to use. In all cases, the cross-section of the element at the free-edge must be square.

The above procedure to obtain the mesh for use in analysis is analogous to the concepts of absolute and conditional convergence in functional analysis [11]. The above procedure uses conditional convergence to establish when a numerical solution is no longer physically consistent or if drifting to an incorrect limit is imminent.

The method produces a valid mesh for analysis by identifying numerical pathologies and not allowing for mesh refinement in their presence (imminent physical inconsistencies and solution drifting). The result is a derived mesh that approximates the exact solution.

With de-homogenization and classical convergence procedures established the topic of implementing the approach of writing the MOS using only critical material properties of the composite constitutive materials can be addressed.

Since this paper was submitted, a more elegant solution to mesh dependency has been developed. Here, an assigned error is not needed, and the target mesh is defined by the onset of divergence in any of the $2n$ strain invariants (n being the number of constituent materials within the composite system). Those interested in this new method should contact the authors.

4 The Use of the Critical Properties to Write the MOS

So far, the basic requirements for implementing a physics-based approach to writing MOS for composite structure have been identified; use only the critical material properties of the constitutive materials of the composite system, derive converged meshes when using numerical methods such as the finite element method and de-homogenize the strain tensors prior to use in the MOS. Two topics remain; how to obtain the critical material properties and how to use them in writing the MOS. Each topic is beyond the scope of this paper therefore, a brief discussion of each will follow. Those interested in a deeper knowledge of these topics should contact the authors.

4.1 Measuring the Critical Properties

Uniaxially tensile loaded unidirectional un-notched laminates are generally used to numerically extract the critical properties of the constitutive material properties of the composite system. Uni-directional laminate coupons consist of one ply orientation and require significant care in their fabrication and testing. If the loading direction is defined as the zero-degree fiber direction, then the coupon with a zero-degree fiber orientation is analyzed at catastrophic failure to numerically extract (with de-homogenization) the critical fiber distortional strain invariant. The coupon with a ninety degree-degree fiber direction is analyzed at catastrophic failure to numerically extract (with de-homogenization) the critical matrix dilatational strain invariant and finally, a coupon with an off-axis fiber orientation (ten to twenty-degrees) is analyzed at catastrophic failure to numerically extract (with de-homogenization) the critical matrix distortional strain invariant. All meshes are converged as discussed in Sect. 3. Specific procedures needed to ensure proper laminate quality and appropriate testing methods can be made available by contacting the authors. These procedures and methods are not standard, and they are needed to establish that the properties extracted are the true material properties and not an artifact of poor fabrication and inappropriate test methods.

4.2 Implementing the Critical Properties into the MOS Process

Once the homogenized strain tensors from a derived mesh have been de-homogenized, the resulting in-situ strain invariants can be calculated, and their values compared to the critical material properties of the constitutive materials. In this paper we are concerned with composite structural details that are subject to complex loading conditions (bi-axial, both in-plane and transverse shear loads). This definition holds for both static and dynamic loading conditions. Structural details will have been loaded as a function of a given load case the global structure must resist (of which there are many). To write the MOS there are three material reference states (assuming a composite system with two constituent materials) within the structural detail that need to be identified for a given loading condition; two for the matrix and one for the fiber.

With respect to complex loading conditions maximal fiber strains within each ply will be either tensile or compressive. Collectively, there will plies with maximal tensile fiber strains and other plies with maximal compressive fiber strains. Of those plies with maximal tensile fiber strains the ply with the greatest maximal matrix distortional strain invariant is scaled until this maximal value equals the critical matrix distortional strain invariant. Similarly, for those plies with maximal fiber compressive strains, the ply with the maximal fiber distortional strain invariant is scaled until this value is equal to the critical fiber distortional strain invariant. These operations result in two different sets of gauss point strain tensors for the same structural detail. Finally, the ply with the maximal dilatational strain invariant is scaled to the critical dilatational strain invariant. These three sets of strains can then be used with (1), or some form of (1), to write the MOS based on the material reference states (relative to the complex loading condition of the current load case).

5 Implementation of the Generalized Finite Element Method to Address Global Structural Assessments

Although submodels can be built to accept the local displacement boundary conditions of the large global model (a vehicle, a device, etc.) to write MOS, the large models are subject to many loads with many more structural details that need MOS assessments. It would be desirable that coarse loads models be the source of the information needed to write the MOS of the structural details of the global structure. Loads models have very little detail, just enough to determine the local loads for structural analysis. If finite elements at the locations of the structural details could be replaced with generalized finite elements representing the structural detail in 3D, then the loads model could be used to determine local loads and the MOS of the structural details with the same efficiency as standard loads models [12]. This concept is currently being investigated at the Fraunhofer Institute.

6 Concluding Remarks

The writing of MOS for composite structure has been based on the use of the building block approach for several decades. Conversely, the writing of MOS for metallic structure has taken a more fundamental approach by using the intrinsic critical material properties of the constitutive materials. In this paper, it has been proposed that it is now possible to use a more fundamental approach to writing MOS for composite structure as well. Issues such as classical convergence, de-homogenization and the use of generalized finite elements have been introduced and discussed (although topics such as Sects. 4.1, 4.2 and 5 are beyond the scope of this paper and require the reader to contact the authors for additional details). The bottom line is, the use of advanced physics-based methods and advanced analysis will allow for more efficient use of composite material systems in structural design as well as allow for a practical approach towards expanding the design space, goals that cannot be realized with the standard building block approach.

References

1. D.J. Peery, J.J. Azar, *Aircraft Structures* (1982). ISBN: 0-07-049196-8
2. American Institute of Steel Construction (AISC), *Manual of Steel Construction*, 8th edn. (American Institute of Steel Construction, Chicago, 1980)
3. R.G. Budynas, J.K. Nisbett, *Mechanical Engineering Design*, 10th edn. (2015). ISBN-13: 978-0-07-339820-9
4. W.F. Chen, D.J. Han, *Plasticity for Structural Engineers* (1988). ISBN: 0-387-96711-7
5. A.A. Shabana, *Computational Continuum Mechanics* (2008). ISBN: 978-0-521-88569-0
6. Y.-L. Shen, *Constrained Deformation of Materials* (2010). ISBN: 978-1-4419-6311-6
7. A. Ritchey, J. Dustin, J. Gosse, R.B. Pipes, *Advances in Composites—Ecodesign and Analysis* (Intech, 2011). ISBN: 978-953-3017-150-3
8. R.J. Jones, *Mechanics of Composite Materials* (1975). ISBN: 0-89116-490-1
9. N.J. Pagano, R.B. Pipes, Some observations on the interlaminar strength of composite laminates, in *Mechanics of Composite Materials, Selected Works of Nicholas J. Pagano*, ed. by J.N. Reddy (Springer, Berlin, 1972). ISBN: 0-7923-3041-2
10. D.S. Burnett, *Finite Element Analysis, from Concepts to Applications* (Addison-Wesley, Reading, 1987). ISBN: 0-201-10806-2
11. F. Morgan, *Real Analysis* (American Mathematical Society, Providence, 2005). ISBN: 0-8218-3670-6
12. M.A. Schweitzer, Multilevel partition of unity method for elliptic problems with strongly discontinuous coefficients, in *Meshfree Methods for Partial Differential Equations VI*, ed. by M. Griebel, M.A. Schweitzer (Springer, Berlin, 2013). ISBN: 978-3-642-32978-4

Kernel-Based Reconstructions for Parametric PDEs

Rüdiger Kempf, Holger Wendland, and Christian Rieger

Abstract In uncertainty quantification, an unknown quantity has to be reconstructed which depends typically on the solution of a partial differential equation. This partial differential equation itself may depend on parameters, some of them may be deterministic and some are random. To approximate the unknown quantity one therefore has to solve the partial differential equation (usually numerically) for several instances of the parameters and then reconstruct the quantity from these simulations. As the number of parameters may be large, this becomes a high-dimensional reconstruction problem.

We will address the topic of reconstructing such unknown quantities using kernel-based reconstruction methods on sparse grids. First, we will introduce into the topic, then explain the reconstruction process and finally provide new error estimates.

1 Introduction

In modern applied sciences dynamic processes are often modeled by partial differential equations, whereby coefficient functions, representing certain material parameters, and forcing terms serve as input. Often, these are obtained by certain measurements or experiments and therefore are prone to being either inaccurate or incomplete and consequently introduce an uncertainty to the model. For a general overview on the topic, see, for example, the recent books [8–11].

R. Kempf (✉) · H. Wendland
Applied and Numerical Analysis, Department of Mathematics, University of Bayreuth, Bayreuth, Germany
e-mail: Ruediger.Kempf@uni-bayreuth.de; Holger.Wendland@uni-bayreuth.de

C. Rieger
Institut für Numerische Simulation, Bonn, Germany
e-mail: rieger@ins.uni-bonn.de

M. Griebel, M. A. Schweitzer (eds.), *Meshfree Methods for Partial Differential Equations IX*, Lecture Notes in Computational Science and Engineering 129,
https://doi.org/10.1007/978-3-030-15119-5_4

In this paper we construct a general framework for the solution of partial differential equations with parametric coefficients. Applications could for example come from problems in groundwater flow, heat transfer or electric fields. To illustrate the general approach to such problems including randomness, we follow [1, 2, 5] and hence restrict ourselves to a Dirichlet-Poisson problem where the parametric diffusion coefficient is given by a function $a : R_{N_P} \times \mathcal{D} \to \mathbb{R}$. The set $R_{N_P} \subset \mathbb{R}^{N_P}$ serves as a finite dimensional parameter space and is, for the sake of simplicity, the hyper-cube $R_{N_P} := \bigtimes_{j=1}^{N_P} \left(-r_j, r_j\right) \subset (-1, 1)^{N_P}$. The number N_P determines the dimension of the parameter space and will be large but finite, i.e. $1 \ll N_P < \infty$, which is known in the literature as *finite noise assumption*.

The parametric partial differential equation is now given on a sufficiently regular domain $\mathcal{D} \subset \mathbb{R}^d$ and for $G \in L^2(\mathcal{D})$ by

$$\begin{aligned} -\nabla \cdot (a(\boldsymbol{y}, \boldsymbol{x}) \nabla u(\boldsymbol{y}, \boldsymbol{x})) &= G(\boldsymbol{x}) \quad \text{in } R_{N_P} \times \mathcal{D}, \\ u(\boldsymbol{y}, \boldsymbol{x}) &= 0 \qquad\quad \text{in } R_{N_P} \times \partial\mathcal{D}, \end{aligned} \tag{1}$$

giving rise to a solution $u : R_{N_P} \times \overline{\mathcal{D}} \to \mathbb{R}$. Obviously, the spatial derivatives are only taken with respect to the spatial variable $\boldsymbol{x}$.

Depending on the practical application, we are not interested in the solution u directly but rather in a derived quantity of interest, which will be modeled by a linear functional q acting on the solution space, i.e.

$$Q(\boldsymbol{y}) := q\left(u(\boldsymbol{y}, \cdot)\right) \in \mathbb{R}, \quad \boldsymbol{y} \in R_{N_P}. \tag{2}$$

Hence, $Q : R_{N_P} \to \mathbb{R}$ is a function operating only on the parameter space. The main task is now to reconstruct the map Q from sampled data $\{Q(\boldsymbol{y}_k)\}$ at specific parameter values $\boldsymbol{y}_k \in \mathbb{Y}_{N_S} := \{\boldsymbol{y}_1, \ldots, \boldsymbol{y}_{N_S}\} \subset R_{N_P}$, $1 \le k \le N_S$, where from now on we denote the number of sampling points by $N_S \in \mathbb{N}$. To avoid any confusion, we note here that N_P and N_S are uncorrelated.

By inserting the sampling points $\boldsymbol{y}_k \in \mathbb{Y}_{N_S}$ into (1), solving the now deterministic Poisson-Dirichlet problem and applying the functional q, we obtain the values $\{Q(\boldsymbol{y}_k)\}$. Except for only very few cases, these steps cannot be done analytically but only numerically. Hence, we introduce a finite dimensional *finite element space* $\mathbb{V}_h \subset \mathbb{V} = H_0^1(\mathcal{D})$, over which we solve

$$\begin{aligned} -\nabla \cdot \left(a(\boldsymbol{y}_k, \boldsymbol{x}) \nabla u_h(\boldsymbol{y}_k, \boldsymbol{x})\right) &= G(\boldsymbol{x}) \quad \text{in } \mathcal{D}, \\ u_h(\boldsymbol{y}_k, \boldsymbol{x}) &= 0 \qquad\quad \text{on } \partial\mathcal{D}, \end{aligned} \tag{3}$$

weakly, yielding an approximation $u_h(\boldsymbol{y}_k, \cdot) \in \mathbb{V}_h$ to the true solution $u(\boldsymbol{y}_k, \cdot) \in \mathbb{V}$ and consequently perturbed samples

$$Q_h(\boldsymbol{y}_k) = q\left(u_h(\boldsymbol{y}_k, \cdot)\right) \approx Q(\boldsymbol{y}_k) = q\left(u(\boldsymbol{y}_k, \cdot)\right). \tag{4}$$

Note, that we assume that we can compute q exactly, i.e., without introducing an additional numerical error. This assumption is made for simplicity and has no impact on the error estimates in Sect. 5.

Choosing a standard finite element method (FEM) for solving (3) weakly, yields well-known error estimates (see for example Brenner and Scott [3]) for the quantities

$$\varepsilon_k := \left\| u_h(\boldsymbol{y}_k, \cdot) - u(\boldsymbol{y}_k, \cdot) \right\|_{\mathbb{V}} \quad \text{and} \quad \varepsilon := \max_{\boldsymbol{y}_k \in \mathbb{Y}_{N_S}} \varepsilon_k. \tag{5}$$

In this paper, the FE method serves as a means to obtain the approximate solution of (1) and can be exchanged with any other suitable PDE solver, in particular also a meshfree solver, as long as there is a known estimate for ε_k.

As mentioned above, we have to choose a discrete sampling set $\mathbb{Y}_{N_S}$ in the high-dimensional space R_{N_P}. This set should on the one hand be dense enough to represent R_{N_P} well and to allow a good reproduction of Q, but on the other hand, since we need to solve a partial differential equation for each element of $\mathbb{Y}_{N_S}$, has to be sparse enough for our method to be applicable. Hence, a natural choice for our set $\mathbb{Y}_{N_S}$ is a *sparse grid* $\mathbb{Y}_{N_S} := H(\ell, N_P)$ of level ℓ in N_P dimensions. However, as our reconstruction technique is purely meshfree other choices are also possible.

The final task is then to reconstruct the high-dimensional function Q from the data $\{(\boldsymbol{y}_k, Q(\boldsymbol{y}_k))\}$ which carry an intrinsic error. Thus, we do not want to use an interpolatory approach but rather a process from standard spline theory, called *smoothing splines* or *penalised least-squares*, see for example [12] and the references therein. The basic structure is a variational problem of the form

$$\widetilde{Q}_\lambda = \arg\min_{s \in \mathcal{H}_K} \sum_{k=1}^{N_S} \left| Q_h(\boldsymbol{y}_k) - s(\boldsymbol{y}_k) \right|^2 + \lambda \|s\|^2_{\mathcal{H}_K}, \tag{6}$$

where $\mathcal{H}_K$ denotes a *reproducing kernel Hilbert space (RKHS)* of real-valued functions with kernel K and where $\lambda > 0$ denotes a penalising parameter.

With this set-up we are able to base our error analysis on a new sampling inequality developed by Wendland and Rieger [6]. The main contribution of this paper, after choosing a specific RKHS $\mathcal{H}_{K_c}$, is the error estimate

$$\begin{aligned} \|Q - \widetilde{Q}_\lambda\|_{L^\infty(R_{N_P})} &\le C\left(f_{N_P,k}(N_S) + \sqrt{\lambda} g_{N_P}(N_S)\right) \|Q\|_{\mathcal{H}_{K_c}} \\ &+ C\left(\frac{1}{\sqrt{\lambda}} f_{N_P,k}(N_S) + g_{N_P}(N_S)\right) N_S h^t \max_{1 \le k \le N_S} \left| u(\boldsymbol{y}_k, \cdot) \right|_{H^{t+1}(\mathcal{D})} \end{aligned}$$

where $C = C(\ell, N_P, k, q) > 0$ is a constant, $\widetilde{Q}_\lambda$ is given in (6), $Q \in \mathcal{H}_{K_c}$ is the function from (2) defined by applying the functional q to the exact solution u of (1), h is the discretisation parameter of the finite element mesh, $t \ge 1$ and $f_{N_P,k}$ and g_{N_P} are functions with known asymptotic behaviour, which are usually derived via so-called sampling inequalites. Examples for the behaviour of $f_{N_P,k}$ and g_{N_P} will be given in Theorem 1. Furthermore, we derive conditions for λ and h such that

$$\|Q - \widetilde{Q}_\lambda\|_{L^\infty(R_{N_P})} \to 0, \quad N_S \to \infty,$$

with the order of $f_{N_P,k}$ and by further sharpening the estimate we even get nearly spectral convergence of the error.

As mentioned above, the setup described above closely follows in particular Griebel and Rieger [5]. However, there are two significant differences to their approach. On the one hand, we use a sparse grid as the sampling set $\mathbb{Y}_{N_S} \subset \mathbb{R}^{N_P}$, instead of a quasi-uniform data set as it is done in [5]. In the given, particular setting this is of significance as we deal with a high-dimensional problem and choosing a sparse grid helps to reduce the effect of the curse of dimensionality. On the other hand, we use a penalised least-squares approach for reconstructing the function Q instead of a support vector machine with Vapnik's loss function, as it has been done in [5]. However, the analysis carried out here for the penalised least-squares problem can easily be replaced by a similar analysis for a support vector machine.

This paper is organised as follows. In Sect. 2 we will review the existence, uniqueness and regularity of the solution of parametric partial differential equations of type (1). In Sect. 3 we will briefly review the theory of reproducing kernel Hilbert spaces (RKHSs), introduce the kernel and associated Hilbert space which we will use throughout this paper and discuss the advantages of the penalised least-squares reconstruction process. A justification for choosing sparse grids as our sampling space $\mathbb{Y}_{N_S}$ will be given in Sect. 4. Finally, in Sect. 5, we will state our main result, the above mentioned error estimate on the reconstruction process, which will be based upon a recently introduced sampling inequality for sparse grids.

We will use the following notation. We denote multi-indices by small, bold Greek letters, e.g. $\boldsymbol{\nu} \in \mathbb{N}_0^d$, and set $\boldsymbol{\nu}! = \prod_{j=1}^d \nu_j!$ and $\boldsymbol{\nu}^{\boldsymbol{\alpha}} = \prod_{j=1}^d \nu_j{}^{\alpha_j}$ for $\boldsymbol{\alpha} \in \mathbb{N}_0^d$. Furthermore, we use the notation $\|\boldsymbol{\nu}\|_1 = \nu_1 + \cdots + \nu_d$ for $\boldsymbol{\nu} \in \mathbb{N}_0^d$.

Additionally, we will use two kinds of Sobolev spaces over domains $\Omega \subseteq \mathbb{R}^d$ of the form $\Omega = \Omega_1 \times \cdots \times \Omega_d$ with $\Omega_j = (-1, 1)$ or $\Omega_j = (-r_j, r_j) \subset (-1, 1)$. On the one hand we will employ the classical Sobolev space

$$W_1^{k,2}(\Omega) := \left\{ f \in L^2(\Omega) \ : \ D^{\boldsymbol{\alpha}} f \in L^2(\Omega), \ \|\boldsymbol{\alpha}\|_1 \le k \right\}$$

equipped with norm

$$\|f\|^2_{W_1^{k,2}(\Omega)} := \sum_{\|\boldsymbol{\alpha}\|_1 \le k} \left\| D^{\boldsymbol{\alpha}} f \right\|^2_{L^2(\Omega)} .$$

On the other hand, we will use the tensor product Sobolev space defined by

$$
\begin{aligned}
W_{\infty}^{k,2}(\Omega) &:= \bigotimes_{j=1}^{d} W^{k,2}(\Omega_j) \\
&= \left\{ f \in L^2(\Omega) \;:\; D^{\boldsymbol{\alpha}} f \in L^2(\Omega),\; \|\boldsymbol{\alpha}\|_\infty \le k \right\}
\end{aligned}
$$

together with the norm

$$
\|f\|^2_{W_{\infty}^{k,2}(\Omega)} := \sum_{\|\boldsymbol{\alpha}\|_\infty \le k} \left\| D^{\boldsymbol{\alpha}} f \right\|^2_{L^2(\Omega)} .
$$

2 Parametric Partial Differential Equations

In this section, we will give an introduction to the theory of parametric partial differential equations by looking at existence and uniqueness of the solutions of the model problem (1). To this end, we need two domains. On the one hand we require a high-dimensional parameter domain. In our case this will be the anisotropic hyper-cube

$$
R_{N_P} := \bigtimes_{j=1}^{N_P} \left(-r_j, r_j\right) \subset (-1,1)^{N_P} , \tag{7}
$$

where $1 \ll N_P < \infty$.

On the other hand we need a spatial domain $\mathcal{D} \subset \mathbb{R}^d$, where usually $d = 2, 3$. We will assume $\mathcal{D}$ to be a bounded, convex and polygonal domain. If $G \in L^2(\mathcal{D})$ then the usual elliptic regularity theory holds for the weak formulation of (1), which is, with the usual energy space $\mathbb{V} := H_0^1(\mathcal{D})$, given by

$$
\int_{\mathcal{D}} a(\boldsymbol{y},\boldsymbol{x}) \nabla u(\boldsymbol{y},\boldsymbol{x}) \cdot \nabla v(\boldsymbol{x})\, d\boldsymbol{x} = \int_{\mathcal{D}} G(\boldsymbol{x}) v(\boldsymbol{x})\, d\boldsymbol{x}, \qquad v \in \mathbb{V},\, \boldsymbol{y} \in R_{N_P}. \tag{8}
$$

In this paper, we assume the coefficient function a to have the form

$$
a(\boldsymbol{y},\boldsymbol{x}) = a_0(\boldsymbol{x}) + \sum_{k=1}^{N_P} \phi_k(\boldsymbol{x}) \boldsymbol{y}_k \tag{9}
$$

with given $\phi_k \in L^\infty(\mathcal{D})$, $k \in \mathbb{N}$, and bounded a_0. In general, the coefficient function a is obtained by a Karhunen-Loève or polynomial chaos expansion which then has to be truncated to give the form (9), so that the restriction to the first N_P terms introduces an additional error, which we will ignore throughout this paper.

We now follow [4] and extend the usual Lax-Milgram theory to complex valued coefficient functions $\tilde{a} : R_{N_P} \times \mathcal{D} \to \mathbb{C}$. To this end, we introduce the so-called *uniform (complex) ellipticity assumption* which requires the existence of two constants $R \geq r > 0$ such that

$$0 < r \leq \Re\left(\tilde{a}(\boldsymbol{y}, \boldsymbol{x})\right) \leq |\tilde{a}(\boldsymbol{y}, \boldsymbol{x})| \leq R \quad \boldsymbol{x} \in \mathcal{D}, \boldsymbol{y} \in R_{N_P}. \tag{10}$$

Here, $\Re(\cdot)$ denotes the real part of a complex number. By rearranging, we see that (10) is satisfied for the function a from (9) if the bounds

$$\sum_{k=1}^{N_P} |\phi_k(\boldsymbol{x})| \leq \Re\left(\min\left(a_0(\boldsymbol{x}) - r, R - a_0(\boldsymbol{x})\right)\right)$$

hold. Here, we have used that $\boldsymbol{y} \in R_{N_P} \subset (-1, 1)^{N_P}$. While this assumption already leads to solutions $u(\boldsymbol{y}, \boldsymbol{x})$, which are analytic as functions of $\boldsymbol{y}$, we need one additional assumption to also bound the coefficients of a Taylor expansion of the function $\boldsymbol{y} \mapsto u(\boldsymbol{y}, \boldsymbol{x})$. Following [4] again, we will call a sequence (ρ_k) of positive numbers δ*-admissible* for the sequence (ϕ_k) if

$$\sum_{k=1}^{\infty} \rho_k |\phi_k(\boldsymbol{x})| \leq \Re\left(a_0(\boldsymbol{x})\right) - \delta, \tag{11}$$

For $\delta \leq r$, one can even have $\rho_k \geq 1$ for all $k \in \mathbb{N}$, see [4]. With this, we have the following result from [4, Theorem 1.2, Lemma 2.4].

Proposition 1 *Suppose that the uniform (complex) ellipticity assumption* (10) *holds with parameters* $0 < r \leq R < \infty$. *Then, the solution of* (8) *has the form*

$$u(\boldsymbol{y}, \cdot) = \sum_{\boldsymbol{\nu} \in \mathbb{N}^{N_P}} u_{\boldsymbol{\nu}}(\cdot) \boldsymbol{y}^{\boldsymbol{\nu}}. \tag{12}$$

for $u_{\boldsymbol{\nu}} \in \mathbb{V}$, *where convergence of the infinite series is understood with respect to the* $\|\cdot\|_{\mathbb{V}}$*-norm. Furthermore, if* (ρ_k) *is a* δ*-admissible sequence, then*

$$\|u_{\boldsymbol{\nu}}(\boldsymbol{y}, \cdot)\|_{\mathbb{V}} \leq \frac{\|G\|_{\mathbb{V}^*}}{\delta} \prod_{k=1}^{N_P} \rho_k^{-\nu_k}, \qquad \boldsymbol{y} \in R_{N_P}.$$

We now want to derive a parametric representation of the quantity of interest (2). To this end, we introduce the notation $\mathcal{R}(\lambda) \in \mathbb{V}$ for the Riesz representer of a linear functional $\lambda \in \mathbb{V}^*$. Then we have, by (12),

$$Q(\boldsymbol{y}) = q\left(u(\boldsymbol{y}, \cdot)\right) = \left\langle \sum_{\boldsymbol{\nu} \in \mathbb{N}^{N_P}} u_{\boldsymbol{\nu}}(\cdot) \boldsymbol{y}^{\boldsymbol{\nu}}, \mathcal{R}(q) \right\rangle_{\mathbb{V}} = \sum_{\nu \in \mathbb{N}^{N_P}} \langle u_{\boldsymbol{\nu}}, \mathcal{R}(q) \rangle_{\mathbb{V}}\, \boldsymbol{y}^{\boldsymbol{\nu}},$$

which shows that the function Q, under certain assumption on the functional q, is also analytic. Later on, this representation of Q will guarantee that Q is indeed an element of the reproducing kernel Hilbert space of our specific choice.

3 Reproducing Kernel Hilbert Spaces

The reconstruction problem (6) is at first sight an optimisation problem over an infinite dimensional function space. However, basic linear algebra shows that the solution must be contained in the span of the Riesz representers of the point evaluation functionals δ_{y_k}. These Riesz representers become particularly simple if the underlying Hilbert space $\mathcal{H}_K$ is a *Hilbert space with a reproducing kernel*. The Hilbert space $\mathcal{H}_K$ is a reproducing kernel Hilbert space with kernel $K : \Omega \times \Omega \to \mathbb{R}$ if K satisfies $K(\cdot, \boldsymbol{y}) \in \mathcal{H}_K$ and $f(\boldsymbol{y}) = \langle f, K(\cdot, \boldsymbol{y})\rangle_{\mathcal{H}_K}$ for all $\boldsymbol{y} \in \Omega$ and all $f \in \mathcal{H}_K$. Details on such spaces can be found, for example, in [13].

3.1 Taylor Spaces and Power Series Kernels

In this paper, we are interested in a particular reproducing kernel Hilbert space, which consists of analytic functions and which was introduced in [14] and further investigated in [15]. The results below are taken from [5].

Let $\boldsymbol{\nu} \in \mathbb{N}_0^{N_P}$ be a multi-index and $(w_{\boldsymbol{\nu}})$ be a sequence of positive numbers such that the summability condition $\sum_{\boldsymbol{\nu}\in\mathbb{N}_0^{N_P}} \frac{w_{\boldsymbol{\nu}}}{\boldsymbol{\nu}!^2} < \infty$ holds. Under these assumptions, a *power series kernel* $K : (-1,1)^{N_P} \times (-1,1)^{N_P} \to \mathbb{R}$, which is a kernel of the form

$$K(\boldsymbol{x}, \boldsymbol{y}) := \sum_{\boldsymbol{\nu}\in\mathbb{N}_0^{N_P}} \frac{w_{\boldsymbol{\nu}}}{\boldsymbol{\nu}!^2} \boldsymbol{x}^{\boldsymbol{\nu}} \boldsymbol{y}^{\boldsymbol{\nu}}, \qquad \boldsymbol{x}, \boldsymbol{y} \in (-1,1)^{N_P}, \tag{13}$$

is well-defined and analytic in each variable. The so-defined kernel K is the reproducing kernel of the Hilbert space $\mathcal{H}_K$ of functions

$$\mathcal{H}_K := \left\{ f : (-1,1)^{N_P} \to \mathbb{R} : f(\boldsymbol{x}) = \sum_{\boldsymbol{\nu}\in\mathbb{N}_0^{N_P}} f_{\boldsymbol{\nu}} \boldsymbol{x}^{\boldsymbol{\nu}} \text{ with } \|f\|_{\mathcal{H}_K} < \infty \right\}, \tag{14}$$

where the norm is defined by the inner product

$$\langle f, g\rangle_{\mathcal{H}_K} := \sum_{\boldsymbol{\nu}\in\mathbb{N}_0^{N_P}} \frac{1}{w_{\boldsymbol{\nu}}} D^{\boldsymbol{\nu}} f(\mathbf{0}) D^{\boldsymbol{\nu}} g(\mathbf{0}) = \sum_{\boldsymbol{\nu}\in\mathbb{N}_0^{N_P}} \frac{\boldsymbol{\nu}!^2}{w_{\boldsymbol{\nu}}} f_{\boldsymbol{\nu}} g_{\boldsymbol{\nu}}, \tag{15}$$

see [14]. The next two results are taken from [5] and illustrate the reason for using such Taylor spaces $\mathcal{H}_K$ in this context.

First, we consider the embedding constant of the embedding of $\mathcal{H}_K$ into either of both Sobolev spaces, the classical isotropic space $W_1^{k,2}$ and the tensor product space $W_\infty^{k,2}$, i.e. we investigate the norm of the injection

$$\mathcal{W}_s(k) : \mathcal{H}_K \hookrightarrow W_s^{k,2}(R_{N_P}), \tag{16}$$

for $s \in \{1, \infty\}$, which is given in the next lemma.

Lemma 1 *Let R_{N_P} be defined by* (7) *Let $s \in \{1, \infty\}$. Suppose that there is a constant $\widehat{c} \in (0, 1)$ such that the weights $w_{\boldsymbol{\nu}}$ satisfy $w_{\boldsymbol{\nu}} \le \widehat{c}^{\|\boldsymbol{\nu}\|_1} \boldsymbol{\nu}!^2$ for all $\boldsymbol{\nu} \in \mathbb{N}_0^{N_P}$. Then, there is a constant $C > 0$ such that the norm of the embedding operator (16) can be bounded by*

$$\|\mathcal{W}_s(k)\| \le \exp\left(\frac{C}{2}k\right) k!.$$

The second result which we require from [5] states that the function Q which we want to reconstruct indeed belongs to a Taylor space $\mathcal{H}_K$ if the weights $w_{\boldsymbol{\nu}}$ are chosen appropriately.

Lemma 2 *Suppose that the uniform (complex) ellipticity assumption* (10) *holds with parameters $0 < r \le R < \infty$. Furthermore, let $(\rho_k)_k$ be a δ-admissible sequence with $0 < \delta < r$ and $\rho_k > 1$ for all k. Let $\mathbf{c} \in \mathbb{R}^{N_P}$ have components c_j with $c_j \in (\rho_j^{-1}, 1)$, $1 \le j \le N_P$. Let $K = K_{\mathbf{c}}$ be defined by (13) with weights $w_{\boldsymbol{\nu}} = \mathbf{c}^{\boldsymbol{\nu}} \boldsymbol{\nu}!^2$. Then we have $Q \in \mathcal{H}_{K_{\mathbf{c}}}$.*

The space $\mathcal{H}_{K_{\mathbf{c}}}$ is a special case of $\mathcal{H}_K$. With the given, specific weights, the inner product becomes

$$\langle f, g\rangle_{\mathcal{H}_{K_{\mathbf{c}}}} = \sum_{\boldsymbol{\nu}\in\mathbb{N}_0^{N_P}} \frac{1}{\mathbf{c}^{\boldsymbol{\nu}} \boldsymbol{\nu}!^2} D^{\boldsymbol{\nu}} f(\mathbf{0}) D^{\boldsymbol{\nu}} g(\mathbf{0}) = \sum_{\boldsymbol{\nu}\in\mathbb{N}_0^{N_P}} \frac{1}{\mathbf{c}^{\boldsymbol{\nu}}} f_{\boldsymbol{\nu}} g_{\boldsymbol{\nu}}.$$

Furthermore, it is easy to see that this specific choice of weights leads to an explicit, analytic form of the reproducing kernel $K_{\mathbf{c}}$ given by

$$K_{\mathbf{c}}(\boldsymbol{x}, \boldsymbol{y}) = \sum_{\boldsymbol{\nu}\in\mathbb{N}_0^{N_P}} \mathbf{c}^{\boldsymbol{\nu}} \boldsymbol{x}^{\boldsymbol{\nu}} \boldsymbol{y}^{\boldsymbol{\nu}} = \prod_{k=1}^{N_P} \frac{1}{1 - c_k x_k y_k}. \tag{17}$$

3.2 *Penalised Least Squares*

A typical application of reproducing kernel Hilbert spaces $\mathcal{H}_K$ are reconstruction processes of the form

$$\min_{s \in \mathcal{H}_K} \left(\sum_{k=1}^{N} |f(\boldsymbol{x}_k) - s(\boldsymbol{x}_k)|^2 + \lambda \|s\|_{\mathcal{H}_K}^2 \right), \tag{18}$$

where the given data $\{(\boldsymbol{x}_k, f(\boldsymbol{x}_k))\}_{1 \le k \le N}$ consist of N samples, which are assumed to come from an otherwise unknown function $f \in \mathcal{H}_K$. The parameter $\lambda > 0$ serves as a moderator between the fit to the data and the smoothness of the reconstruction $\widetilde{s}_\lambda$. In the RKHS setting, we have, by the well-known representer theorem, that the solution of the minimisation $\widetilde{s}_\lambda$ lies in the finite-dimensional space spanned by $K(\cdot, \boldsymbol{x}_k)$, $1 \le k \le N$, i.e. we have the representation

$$\widetilde{s}_\lambda = \sum_{k=1}^{N} \alpha_k K(\cdot, \boldsymbol{x}_k).$$

Furthermore, the coefficients $\boldsymbol{\alpha} = (\alpha_1, \ldots, \alpha_N)^T$ can be computed by solving the linear system

$$(\boldsymbol{K} + \lambda \boldsymbol{I})\boldsymbol{\alpha} = \boldsymbol{f},$$

where $\boldsymbol{K}_{ij} = K(\boldsymbol{x}_i, \boldsymbol{x}_j)$, $\boldsymbol{f}_i = f(\boldsymbol{x}_i)$ and $\boldsymbol{I}$ is the identity matrix. It is well-known that this system has a positive definite system matrix and hence a unique solution. This also means that the least-squares problem (18) has a unique solution. Hence, in our situation, when employing the kernel $K = K_c$ in (6), these general considerations give us a unique approximation $\widetilde{Q}_\lambda$ to Q derived from the noisy data $Q_h(\boldsymbol{y}_k)$, $1 \le k \le N_S$.

4 Sparse Grids

In this section, we demonstrate how we construct the sparse grid $H(\ell, d)$ of level ℓ and dimension d. Here, we follow mainly Wendland and Rieger [6].

To obtain the d-dimensional grid, we start with univariate sets of Chebyshev points. To do so we define a sequence of numbers (m_i) by

$$\begin{aligned} m_1 &= 1, \\ m_i &= 2^{i-1} + 1, \quad i > 1. \end{aligned}$$

Then, we define the Chebyshev point sets $X^{(i)}$ to be

$$X^{(1)} := X_{m_1} = \{0\},$$
$$X^{(i)} := X_{m_i} = \left\{ x_j^{(i)} = -\cos\left(\frac{\pi(j-1)}{m_i - 1}\right) \;:\; 1 \le j \le m_i \right\}, \qquad i > 1.$$

With these univariate point sets, we now define the *sparse grid* $\widetilde{H}(\ell, d)$ of level ℓ and dimension d, $\ell \ge d$, by

$$\widetilde{H}(\ell, d) = \bigcup_{\substack{\boldsymbol{i}\in\mathbb{N}^d \\ \|\boldsymbol{i}\|_1=\ell}} X^{(i_1)} \times \cdots \times X^{(i_d)}. \tag{19}$$

As mentioned in Sect. 1, we choose the sampling space $\mathbb{Y}_{N_S}$ to be a sparse grid. As, by construction, $\widetilde{H}(\ell, N_P) \subset [-1, 1]^{N_P}$ is not a subset of R_{N_P}, we simply scale its points with a component-wise factor $r_j(1-\mu)$, $0 < \mu \ll 1$, $1 \le j \le N_P$ and receive the *scaled sparse grid*

$$H(\ell, N_P) := \left\{ \left(r_1(1-\mu)x_1, \ldots, r_{N_P}(1-\mu)x_{N_P}\right) \;:\; \boldsymbol{x} \in \widetilde{H}(\ell, N_P) \right\}. \tag{20}$$

Now, we choose $\mathbb{Y}_{N_S} := H(\ell, N_P) \subset R_{N_P}$, where ℓ is a degree of freedom. For statements on the error of the reconstruction process we need to know the number of sampling points N_S. Unfortunately, an explicit formula for this number is unknown and there exist only lower and upper bounds, provided in [6],

$$2^{\ell-2N_P+1} \le N_S \le 2^{\ell-N_P+1}\frac{\ell^{N_P-1}}{(N_P-1)!}.$$

Fortunately, as soon as we have created the sparse grid, we know exactly how many points it contains. A selection is given in Table 1. Clearly, we can control the number N_S for a given dimension N_P by choosing the level ℓ appropriately.

5 Error Estimates

We use this section to state the main results of this paper concerning error estimates for the optimisation problem (6), whose definition we recall here. We use the reproducing kernel Hilbert space $\mathcal{H}_{K_c}$ introduced in (14) with the power series kernel $K_c(\boldsymbol{x}, \boldsymbol{y}) = \prod_{j=1}^{N_P} \frac{1}{1-c_j x_j y_j}$ of (17). Next, we define $J_{Q_h,\lambda} : \mathcal{H}_{K_c} \to \mathbb{R}$, where

$$J_{Q_h,\lambda}(s) := \sum_{k=1}^{N_P} \left|Q_h(\boldsymbol{y}_k) - s(\boldsymbol{y}_k)\right|^2 + \lambda\|s\|^2_{\mathcal{H}_{K_c}}, \qquad s \in \mathcal{H}_{K_c},$$

Table 1 Number of points N_S of the grid $H(\ell, d)$ for various space dimensions d and $\ell \geq d$

$\ell \mid d$	2	3	4	5	6	7	8
2	1						
3	5	1					
4	13	7	1				
5	29	25	9	1			
6	65	69	41	11	1		
7	145	177	137	61	13	1	
8	321	441	401	241	85	15	1
9	705	1073	1105	801	389	113	17
10	1537	2561	2929	2433	1457	589	145
11	3329	6017	7537	6993	4865	2465	849
12	7169	13,953	18,945	19,313	15,121	9017	3937
13	15,361	32,001	46,721	51,713	44,689	30,241	15,713
14	32,769	72,705	113,409	135,073	127,105	95,441	56,737
15	69,633	163,841	271,617	345,665	350,657	287,745	190,881

and set

$$\widetilde{Q}_\lambda := \arg\min_{s \in \mathcal{H}_{K_c}} J_{Q_h,\lambda}(s).$$

The main objective is now to reconstruct the function $Q : R_{N_P} \to \mathbb{R}$ from perturbed samples $Q_h(\boldsymbol{y}_k) = q(u_h(\boldsymbol{y}_k, \cdot))$, $1 \leq k \leq N_S$, where the $\boldsymbol{y}_k \in H(\ell, N_P)$ and $u_h(\boldsymbol{y}_k, \cdot) \in \mathbb{V}_h$ is a FEM approximation of the exact solution.

As the data we have are corrupted by numerical error, we cannot directly employ the classic error estimates for penalised least-squares used in [6] since we cannot assume the function Q_h to be in the Hilbert space $\mathcal{H}_{K_c}$. Nonetheless, we can assess this error. To do so we use the quantity

$$\varepsilon_k = Q_h(\boldsymbol{y}_k) - Q(\boldsymbol{y}_k) = q(u_h(\boldsymbol{y}_k, \cdot)) - q(u(\boldsymbol{y}_k, \cdot)) = q\left(u_h(\boldsymbol{y}_k, \cdot) - u(\boldsymbol{y}_k, \cdot)\right).$$

Hence we have the estimate

$$|\varepsilon_k| \leq \|q\|_{\mathbb{V}^*} \|u_h(\boldsymbol{y}_k, \cdot) - u(\boldsymbol{y}_k, \cdot)\|_{\mathbb{V}}$$

which means that $|\varepsilon_k|$ is bounded by the numerical error, which occurs in the solution of equation (3). This error enjoys well-known bounds depending on the smootheness of the solution. For example, we have from Brenner and Scott [3] the following estimate.

Lemma 3 *Let the finite element space be made up of elements up to order t with mesh width h. Assume that $u(\boldsymbol{y}_k, \cdot) \in H^{s+1}(\mathcal{D})$, for an $1 \le s \le t$ and all $1 \le k \le N_P$. Then, there is a constant $c > 0$ such that*

$$\|u_h(\boldsymbol{y}_k, \cdot) - u(\boldsymbol{y}_k, \cdot)\|_{\mathbb{V}} \le ch^s |u(\boldsymbol{y}_k, \cdot)|_{H^{s+1}(\mathcal{D})}, \quad 1 \le k \le N_P. \tag{21}$$

Another tool we require is a sampling inequality, which allows us to bound the L^∞-norm of a function by a weighted sum of a full Sobolev norm and an ℓ^∞-norm over the discrete sampling set. The particular inequality we use is a new approach tailored for sparse grids. It gives the weights in terms of the number of sampling points and not in terms of the fill distance of the discrete set as it is usually done in sampling inequalities. This is of particular importance when working with sparse grids and in higher dimensions. The version we use in this paper is a special case of the one presented in [6].

Theorem 1 *Let $\widetilde{H}(\ell, N_P)$, $\ell \ge N_P$, be the sparse grid of (19) with N_S points. Then, for every function $a \in W^{k,2}_\infty\left((-1,1)^{N_P}\right)$, $k \in \mathbb{N}$, there exists a constant $\widetilde{c} := \widetilde{c}(\ell, N_P, k)$ such that*

$$\|a\|_{L^\infty((-1,1)^{N_P})} \le \widetilde{c}\left(f_{N_P,k}(N_S)\|a\|_{W^{k,2}_\infty((-1,1)^{N_P})} + g_{N_P}(N_S)\|a\|_{\ell^\infty(\widetilde{H}(\ell,N_P))}\right). \tag{22}$$

Here, the functions $f_{N_P,k}$ and g_{N_P} have for $N_S \to \infty$ the asymptotic behaviour

$$f_{N_P,k}(N_S) = O\left(N_S^{-k+\frac{1}{2}} \log(N_S)^{N_P\left(k+\frac{5}{2}\right)-\left(k+\frac{1}{2}\right)}\right), \tag{23}$$

$$g_{N_P}(N_S) = O\left(\log(N_S)^{N_P}\right). \tag{24}$$

In [6], the weight-functions $f_{N_P,k}, g_{N_P} : \mathbb{N} \to \mathbb{R}$ are given explicitly, but for our purposes the asymptotic behaviour is sufficient. Obviously, the function $f_{N_P,k}$ goes to zero for $N_S \to \infty$ while the function g_{N_P} grows logarithmically.

As (22) holds for the unscaled sparse grid $\widetilde{H}(\ell, N_P) \subset [-1,1]^{N_P}$, we need to modify it to fit into our framework, namely the scaled sparse grid $H(\ell, N_P)$ of (20). We scale the occuring functions by a simple coordinate transform, i.e., we scale the arguments by the same factors we used in the construction of $H(\ell, N_P)$ in Sect. 4. In doing so, we arrive at

$$\|b\|_{L^\infty(R_{N_P})} \le c\left(f_{N_P,k}(N_S)\|b\|_{W^{k,2}_\infty(R_{N_P})} + g_{N_P}(N_S)\|b\|_{\ell^\infty(H(\ell,N_P))}\right), \tag{25}$$

where $b \in W^{k,2}_\infty\left(R_{N_P}\right)$, $f_{N_P,k}$ and g_{N_P} are the functions of Theorem 1 and $c := c(\ell, N_P, k, \mu, \boldsymbol{r})$ is a modified constant, depending additionally on μ and $\boldsymbol{r}$.

As we can embed $\mathcal{H}_{K_c}$ into $W^{k,2}_\infty\left(R_{N_P}\right)$, see Lemma 1, (25) above holds particularly for functions $b \in \mathcal{H}_{K_c}$.

With these tools we are now able to estimate the error $Q - \widetilde{Q}_\lambda$ in the L^∞-norm. We mainly follow the ideas employed in [5] with appropriate modifications. We start by deriving two estimates for $\widetilde{Q}_\lambda$. The first one is based upon the bound

$$\begin{aligned} \left|Q_h(\boldsymbol{y}_k) - \widetilde{Q}_\lambda(\boldsymbol{y}_k)\right|^2 &\leq \sum_{i=1}^{N_S} \left|Q_h(\boldsymbol{y}_i) - \widetilde{Q}_\lambda(\boldsymbol{y}_i)\right|^2 + \lambda \left\|\widetilde{Q}_\lambda\right\|^2_{\mathcal{H}_{K_c}} \\ &= J_{Q_h,\lambda}\left(\widetilde{Q}_\lambda\right) \\ &\leq J_{Q_h,\lambda}(Q), \end{aligned}$$

where we introduced positive summands and used that $\widetilde{Q}_\lambda$ is the minimiser of the functional $J_{Q_h,\lambda}$. This leads to

$$\left|Q_h(\boldsymbol{y}_k) - \widetilde{Q}_\lambda(\boldsymbol{y}_k)\right| \leq \sum_{i=1}^{N_S} \left|Q_h(\boldsymbol{y}_i) - Q(\boldsymbol{y}_i)\right| + \sqrt{\lambda}\, \|Q\|_{\mathcal{H}_{K_c}}. \tag{26}$$

Here, we used that for any $a, b \geq 0$ the estimate $(a+b)^{1/2} \leq a^{1/2} + b^{1/2}$ holds.

Next, we can estimate the consistency error, i.e. the point-wise error at the sampling nodes. We have by applying the triangle inequality

$$\left|Q(\boldsymbol{y}_k) - \widetilde{Q}_\lambda(\boldsymbol{y}_k)\right| \leq \left|Q(\boldsymbol{y}_k) - Q_h(\boldsymbol{y}_k)\right| + \left|Q_h(\boldsymbol{y}_k) - \widetilde{Q}_\lambda(\boldsymbol{y}_k)\right|,$$

which, together with (26), leads to our first crucial estimate

$$\left|Q(\boldsymbol{y}_k) - \widetilde{Q}_\lambda(\boldsymbol{y}_k)\right| \leq \left|Q_h(\boldsymbol{y}_k) - Q(\boldsymbol{y}_k)\right| + \sum_{i=1}^{N_S} \left|Q_h(\boldsymbol{y}_i) - Q(\boldsymbol{y}_i)\right| + \sqrt{\lambda}\, \|Q\|_{\mathcal{H}_{K_c}}.$$

The second estimate on $\widetilde{Q}_\lambda$ follows from

$$\lambda \left\|\widetilde{Q}_\lambda\right\|^2_{\mathcal{H}_{K_c}} \leq J_{Q_h,\lambda}\left(\widetilde{Q}_\lambda\right) \leq J_{Q_h,\lambda}(Q) \leq \sum_{k=1}^{N_S} \left|Q_h(\boldsymbol{y}_k) - Q(\boldsymbol{y}_k)\right|^2 + \lambda\, \|Q\|^2_{\mathcal{H}_{K_c}}$$

and leads to

$$\left\|\widetilde{Q}_\lambda\right\|^2_{\mathcal{H}_{K_c}} \leq \frac{1}{\lambda} \sum_{k=1}^{N_S} \left|Q_h(\boldsymbol{y}_k) - Q(\boldsymbol{y}_k)\right|^2 + \|Q\|^2_{\mathcal{H}_{K_c}}.$$

We collect these results in the following lemma.

Lemma 4 *The reconstruction $\widetilde{Q}_\lambda$ from (6) of the function Q satisfies the bounds*

$$\left|Q(\boldsymbol{y}_k) - \widetilde{Q}_\lambda(\boldsymbol{y}_k)\right| \leq \left|Q_h(\boldsymbol{y}_k) - Q(\boldsymbol{y}_k)\right| + \sum_{i=1}^{N_S} \left|Q_h(\boldsymbol{y}_i) - Q(\boldsymbol{y}_i)\right| + \sqrt{\lambda}\, \|Q\|_{\mathcal{H}_{K_c}},$$

$$\|\widetilde{Q}_\lambda\|^2_{\mathcal{H}_{K_c}} \leq \frac{1}{\lambda} \sum_{i=1}^{N_S} \left|Q_h(\boldsymbol{y}_i) - Q(\boldsymbol{y}_i)\right|^2 + \|Q\|^2_{\mathcal{H}_{K_c}}.$$

With these results we arrive at the following error estimate.

Theorem 2 *Let $H(\ell, N_P)$ with $\ell \geq N_P$ be the scaled sparse grid from* (20) *with N_S points. Assume that $Q \in \mathcal{H}_{K_c}$, where $\mathcal{H}_{K_c}$ is as in* (14) *with $K = K_c$ from (17) satisfying the assumptions of Lemma 2. Then, for each sufficiently large k there is a constant $c := c(\ell, N_P, k, \mu, \boldsymbol{r}) > 0$ such that for $\widetilde{Q}_\lambda = \arg\min_{s \in \mathcal{H}_{K_c}} J_{Q_h,\lambda}(s)$ the error estimate*

$$\begin{aligned}\|Q - \widetilde{Q}_\lambda\|_{L^\infty(R_{N_P})} &\leq c\left(f_{N_P,k}(N_S) + \sqrt{\lambda} g_{N_P}(N_S)\right) \|Q\|_{\mathcal{H}_{K_c}} \\ &\quad + c\left(\frac{1}{\sqrt{\lambda}} f_{N_P,k}(N_S) + g_{N_P}(N_S)\right) \sum_{i=1}^{N_S} \left|Q_h(\boldsymbol{y}_i) - Q(\boldsymbol{y}_i)\right|\end{aligned}$$

holds, where $f_{N_P,k}$ and g_{N_P} are from (23) *and* (24).

Proof The modified sampling inequality (25) with $b = Q - \widetilde{Q}_\lambda$ and Lemma 1 show

$$\begin{aligned}\|Q - \widetilde{Q}_\lambda\|_{L^\infty(R_{N_P})} &\leq c f_{N_P,k}(N_S) \|Q - \widetilde{Q}_\lambda\|_{W^{k,2}_\infty(R_{N_P})} \\ &\quad + c g_{N_P}(N_S) \|Q - \widetilde{Q}_\lambda\|_{\ell^\infty(H(\ell,N_P))} \\ &\leq c f_{N_P,k}(N_S) \|Q - \widetilde{Q}_\lambda\|_{\mathcal{H}_{K_c}} \\ &\quad + c g_{N_P}(N_S) \|Q - \widetilde{Q}_\lambda\|_{\ell^\infty(H(\ell,N_P))}.\end{aligned}$$

Next, Lemma 4 allows us to bound the terms $\|Q - \widetilde{Q}_\lambda\|_{\ell^\infty(H(\ell,N_P))}$ and $\|Q - \widetilde{Q}_\lambda\|_{\mathcal{H}_{K_c}}$ separately. We have

$$\begin{aligned}&\|Q - \widetilde{Q}_\lambda\|_{\ell^\infty(H(\ell,N_P))} \\ &\leq \max_{k=1,\ldots,N_S} \left|Q_h(\boldsymbol{y}_k) - Q(\boldsymbol{y}_k)\right| + \sum_{i=1}^{N_S} \left|Q_h(\boldsymbol{y}_i) - Q(\boldsymbol{y}_i)\right| + \sqrt{\lambda}\, \|Q|_{\mathcal{H}_{K_c}} \\ &\leq 2 \sum_{i=1}^{N_S} \left|Q_h(\boldsymbol{y}_i) - Q(\boldsymbol{y}_i)\right| + \sqrt{\lambda}\, \|Q\|_{\mathcal{H}_{K_c}}\end{aligned}$$

and

$$\left\| Q - \widetilde{Q}_\lambda \right\|_{\mathcal{H}_{K_c}} \leq \|Q\|_{\mathcal{H}_{K_c}} + \left\| \widetilde{Q}_\lambda \right\|_{\mathcal{H}_{K_c}}$$
$$\leq 2\,\|Q\|_{\mathcal{H}_{K_c}} + \frac{1}{\sqrt{\lambda}} \sum_{i=1}^{N_S} \left| Q_h(\mathbf{y}_i) - Q(\mathbf{y}_i) \right|.$$

Inserting these bounds into the above bound on $\left\| Q - \widetilde{Q}_\lambda \right\|_{L^\infty(R_{N_P})}$ concludes the proof. □

Taking also the error bound (21) of the finite element approximation into account yields the following corollary.

Corollary 1 *Let the assumptions of Theorem 2 hold. Assume further that* $u(\mathbf{y}_i, \cdot) \in H^{t+1}(\mathcal{D})$, $t \in \mathbb{N}$, *for every* $\mathbf{y}_i \in H(\ell, N_P)$, $1 \leq i \leq N_S$. *Then the error estimate*

$$\left\| Q - \widetilde{Q}_\lambda \right\|_{L^\infty(R_{N_P})} \leq c(\ell, N_P, k, \mu, \mathbf{r}) \left(f_{N_P,k}(N_S) + \sqrt{\lambda} g_{N_P}(N_S) \right) \|Q\|_{\mathcal{H}_{K_c}}$$
$$+ c(\ell, N_P, k, \mu, \mathbf{r}, q) \left(\frac{1}{\sqrt{\lambda}} f_{N_P,k}(N_S) + g_{N_P}(N_S) \right) N_S h^t \max_{1\leq i \leq N_S} \left| u(\mathbf{y}_i, \cdot) \right|_{H^{t+1}(\mathcal{D})}$$

holds. The functions $f_{N_P,k}$ *and* g_{N_P} *are from* (23) *and* (24).

Note that the constant depends on the smoothness k and that the error bounds are for fixed but arbitrarily large k. Future research should address the precise asymptotic behaviour of this constant with $k \to \infty$.

Next, we want to discuss the convergence behaviour of the estimate above. As it is, this result is problematic since for $N_S \to \infty$ the function g_{N_P} tends to infinity. Hence, to achieve convergence, we have to couple the penalisation parameter λ and the mesh width h of the finite element grid to the number of points N_S in our sparse grid appropriately.

We start with the first term on the right-hand side of the bound in Corollary 1 . Its behaviour is determined by

$$f_{N_P,k}(N_S) + \sqrt{\lambda} g_{N_P}(N_S),$$

To have this term to behave like $f_{N_P,k}(N_S)$, which converges to zero for $N_S \to \infty$, we must choose $\sqrt{\lambda}$ sufficiently small . However, as we have a $1/\sqrt{\lambda}$ in the second term of the bound of Corollary 1, we cannot choose it too small. Hence, we choose a proportional constant $c_p > 0$ and let

$$\lambda = c_p \left(\frac{f_{N_P,k}(N_S)}{g_{N_P}(N_S)} \right)^2. \tag{27}$$

With this choice, the bound in Corollary 1 becomes

$$\|Q - \widetilde{Q}_\lambda\|_{L^\infty(R_{N_P})} \le c\left(f_{N_P,k}(N_S)\|Q\|_{\mathcal{H}_{K_c}} + g_{N_P}(N_S)N_S h^t \max_{1\le k\le N_S}\left|u(\boldsymbol{y}_k,\cdot)\right|_{H^{t+1}(\mathcal{D})}\right).$$

Hence, in order to have convergence, we need to ensure that the factor $g_{N_P}(N_S)N_S h^t$ in the second summand also tends to zero. If we want to keep the convergence order of size $f_{N_P,k}(N_S)$ then we have to choose h to satisfy

$$h \le \left(\frac{f_{N_P,k}(N_S)}{N_S g_{N_P}(N_S)}\right)^{\frac{1}{t}}. \tag{28}$$

We summarise these results in the next corollary.

Corollary 2 *Under the assumptions of Corollary 1 and with the choices (27) for the smoothing parameter and (28) for the finite element mesh size, the reconstruction error satisfies*

$$\|Q - \widetilde{Q}_\lambda\|_{L^\infty(R_{N_P})} \le c f_{N_P,k}(N_S) = c N_S^{-k+\frac{1}{2}} \log(N_S)^{N_P\left(k+\frac{5}{2}\right)-\left(k+\frac{1}{2}\right)}$$

with a constant $c = c(Q, u, k, \ell, N_P, \mu, \boldsymbol{r})$ *depending additionally on* Q *and* u.

Comparing this result to the one obtained in [5] and by experience from classical RKHS results, see [13], one would, in light of the analycity of the kernel, expect spectral convergence of the reconstruction error, similar to [7]. And indeed, a more thorough analysis of the occuring constants leads to the following result.

Corollary 3 *Under the assumptions of Corollary 1 and with the choices (27) for the smoothing parameter and (28) for the finite element mesh size, the reconstruction error satisfies for sufficiently large* N_S *the bound*

$$\|Q - \widetilde{Q}_\lambda\|_{L^\infty(R_{N_P})} \le c_1 {N_S}^2 e^{-c_2 N_S}\left(\|Q\|_{\mathcal{H}_{K_c}} + \max_{1\le k\le N_S}\left|u(\boldsymbol{y}_k,\cdot)\right|_{H^{t+1}(\mathcal{D})}\right),$$

where $c_1, c_2 > 0$ *are constants.*

Proof Corollary 2, together with the embedding constant from Lemma 1, gives the estimate

$$\|Q - \widetilde{Q}_\lambda\|_{L^\infty(R_{N_P})} \le c k! e^{\frac{C}{2}k} N_S^{-k+\frac{1}{2}} \log(N_S)^{N_P\left(k+\frac{5}{2}\right)-\left(k+\frac{1}{2}\right)}\left(\|Q\|_{\mathcal{H}_{K_c}} + \max_{1\le k\le N_S}\left|u(\boldsymbol{y}_k,\cdot)\right|_{H^{t+1}(\mathcal{D})}\right)$$

Using Stirling's estimate $k! \le ck^{k+\frac{1}{2}}e^{-k}$, $k \ge 1$, and keeping in mind that the logarithmic term $\log(N_S)$ grows slower than any root of N_S, especially slower than $N_S{}^{\frac{1}{N_P(k+5/2)-(k+1/2)}}$, we obtain

$$\begin{aligned}
&\|Q - \widetilde{Q}_\lambda\|_{L^\infty(R_{N_P})} \\
&\le ck^{k+\frac{1}{2}}e^{-k}e^{\frac{C}{2}k}N_S^{-k+\frac{1}{2}}N_S\left(\|Q\|_{\mathcal{H}_{K_c}} + \max_{1\le k\le N_S}\left|u(\boldsymbol{y}_k,\cdot)\right|_{H^{t+1}(\mathcal{D})}\right) \\
&= c\left(kN_S{}^3\right)^{\frac{1}{2}}\left(e^{1-\frac{C}{2}}\frac{N_S}{k}\right)^{-k}\left(\|Q\|_{\mathcal{H}_{K_c}} + \max_{1\le k\le N_S}\left|u(\boldsymbol{y}_k,\cdot)\right|_{H^{t+1}(\mathcal{D})}\right). \quad (29)
\end{aligned}$$

Next, for sufficiently large N_S, we choose k as $k = \frac{N_S}{\nu}$, where ν is a fixed constant such that $k \in \mathbb{N}$ and $e^{\frac{C}{2}-1} < \nu \le N_S$ holds. Inserting this particular choice of k into (29) yields

$$\begin{aligned}
&\|Q - \widetilde{Q}_\lambda\|_{L^\infty(R_{N_P})} \\
&\le c\nu^{-\frac{1}{2}}N_S{}^2\left(\nu e^{1-\frac{C}{2}}\right)^{-\frac{N_S}{\nu}}\left(\|Q\|_{\mathcal{H}_{K_c}} + \max_{1\le k\le N_S}\left|u(\boldsymbol{y}_k,\cdot)\right|_{H^{t+1}(\mathcal{D})}\right) \\
&= c\nu^{-\frac{1}{2}}N_S{}^2\left(e^{\frac{1}{\nu}\left(1-\frac{C}{2}+\log\nu\right)}\right)^{-N_S}\left(\|Q\|_{\mathcal{H}_{K_c}} + \max_{1\le k\le N_S}\left|u(\boldsymbol{y}_k,\cdot)\right|_{H^{t+1}(\mathcal{D})}\right) \\
&= c_1N_S{}^2e^{-c_2N_S}\left(\|Q\|_{\mathcal{H}_{K_c}} + \max_{1\le k\le N_S}\left|u(\boldsymbol{y}_k,\cdot)\right|_{H^{t+1}(\mathcal{D})}\right),
\end{aligned}$$

with $c_1 = c(\ell, N_P, \mu, \boldsymbol{r}, q)\nu^{-\frac{1}{2}} > 0$ and $c_2 = \frac{1}{\nu}\left(1 - \frac{C}{2} + \log\nu\right) > 0$ for ν in the given range. □

6 Concluding Remarks and Future Work

We have recaptured the basics of the regularity theory of parametric elliptic partial differential equations. One important result was that the solution, as a function of the parameter, is analytic and hence so is the quantity of interest.

The analyticity of the function we wanted to reconstruct motivated the choice of the specific reproducing kernel Hilbert space, a Taylor space, whose kernel is a power series kernel and thus analytic itself. With these choices we employed a regularised reconstruction process for approximating the smooth function from data which are usually corrupted by a (numerical) error, which means that the data-giving function is not an element of the approximation space.

To alleviate the curse of dimensionality we employed sparse grids, and a new type of sampling inequality which is expressed in the number of points rather than the fill distance of the sampling set.

We used the two degrees of freedom at our disposal, namely the FEM mesh width and the penalty parameter of the reconstruction process, to derive an overall error estimate.

The next step is to verify the derived theoretical results numerically. This will be pursued in the future.

Finally, the kernel we used is globally supported which will lead to dense system matrices which should be avoided in practical applications, especially if the number of sampling points, i.e. the dimension of the matrix, becomes large. Switching to compactly supported kernels is subject of ongoing research. However, due to the high-dimensional nature of the underlying domain, a compactly supported kernel might not "see" enough information unless various scales are employed.

Acknowledgements Christian Rieger would like to thank the Deutsche Forschungsgemeinschaft (DFG) for financial support through the CRC 1060, The Mathematics of Emergent Effects.

References

1. I. Babuska, R. Tempone, G.E. Zouraris, Galerkin finite element approximations of stochastic elliptic partial differential equations. SIAM J. Numer. Anal. **42**, 800–825 (2004)
2. I. Babuska, F. Nobile, R. Tempone, A stochastic collocation method for elliptic partial differential equations with random input data. SIAM J. Numer. Anal. **45**, 1005–1034 (2007)
3. S. Brenner, L.R. Scott, *The Mathematical Theory of Finite Element Methods*. Texts in Applied Mathematics (Springer, New York, 2002)
4. A. Cohen, R. DeVore, C. Schwab, Analytic regularity and polynomial approximation of parametric and stochastic elliptic PDEs. Anal. Appl. **9**(1), 11–47 (2010)
5. M. Griebel, C. Rieger, Reproducing kernel Hilbert spaces for parametric partial differential equations. SIAM/ASA J. Uncertain. Quantif. **5**, 111–137 (2017)
6. C. Rieger, H. Wendland, Sampling inequalities for sparse grids. Numer. Math. **136**, 439–466 (2017)
7. C. Rieger, B. Zwicknagl, Sampling inequalities for infinitely smooth functions, with applications to interpolation and machine learning. Adv. Comput. Math. **32**, 103–129 (2010)
8. R.C. Smith, *Uncertainty Quantification*. Computational Science & Engineering, vol. 12 (Society for Industrial and Applied Mathematics (SIAM), Philadelphia, PA, 2014). Theory, implementation, and applications. MR 3155184
9. C. Soize, *Uncertainty Quantification*, Interdisciplinary Applied Mathematics, vol. 47 (Springer, Cham, 2017). An accelerated course with advanced applications in computational engineering. With a foreword by Charbel Farhat. MR 3618803
10. T.J. Sullivan, *Introduction to Uncertainty Quantification*. Texts in Applied Mathematics, vol. 63 (Springer, Cham, 2015). MR 3364576
11. L. Tenorio, *An Introduction to Data Analysis and Uncertainty Quantification for Inverse Problems*. Mathematics in Industry (Philadelphia) (Society for Industrial and Applied Mathematics (SIAM), Philadelphia, PA, 2017). MR 3672154
12. G. Wahba, Spline Models for Observational Data, CBMS-NSF, Regional Conference Series in Applied Mathematics (SIAM, Philadelphia, 1990)

13. H. Wendland, Scattered Data Approximation (Cambridge University Press, Cambridge, 2004)
14. B. Zwicknagl, Power series kernels. Constr. Approx. **29**(1), 61–84 (2009)
15. B. Zwicknagl, R. Schaback, Interpolation and approximation in taylor spaces. J. Approx. Theory **171**, 65–83 (2013)

Fluid Structure Interaction (FSI) in the MESHFREE Finite Pointset Method (FPM): Theory and Applications

Jörg Kuhnert, Isabel Michel, and Reiner Mack

Abstract Fluid Structure Interaction (FSI) and meshfree numerical methods are a perfect couple. One often repeated strong argument is the almost natural coupling of meshfree methods in a Lagrangian/ALE formulation with moving, flexible structures.

Since 1996, Fraunhofer ITWM has been developing a Generalized Finite Difference Method (GFDM), a purely meshfree solver for fluid and continuum mechanics. In the industrial context, this method is also referred to as Finite Pointset Method (FPM). Currently, it is further developed to an integrated tool called MESHFREE which combines the advantages of GFDM/FPM as well as SAMG, a fast solver for large sparse linear systems developed by Fraunhofer SCAI. This synergy drastically increases the applicability of the method since SAMG provides a robust and scalable linear solver for a wide class of problems.

In this contribution, we classify fundamental FSI aspects in GFDM/FPM: classical pressure–velocity coupling and alternative velocity–pressure coupling. Each category will be illustrated by industrially relevant examples, with special focus on Pelton turbine applications and flow in flexible tubes.

1 Introduction

The Finite Pointset Method (FPM) is a meshfree Generalized Finite Difference Method (GFDM) in fluid and continuum mechanics. It solves partial differential equations directly on a cloud of numerical points, i.e. it uses the strong formulation. By default, the classical conservation equations (mass, momentum, and energy) in

J. Kuhnert (✉) · I. Michel
Fraunhofer Institute for Industrial Mathematics ITWM, Kaiserslautern, Germany
e-mail: joerg.kuhnert@itwm.fraunhofer.de; isabel.michel@itwm.fraunhofer.de

R. Mack
Voith Hydro Holding GmbH & Co. KG, Heidenheim, Germany
e-mail: Reiner.Mack@voith.com

M. Griebel, M. A. Schweitzer (eds.), *Meshfree Methods for Partial Differential Equations IX*, Lecture Notes in Computational Science and Engineering 129,
https://doi.org/10.1007/978-3-030-15119-5_5

combination with general material models are solved. The approximation of spatial partial differential operators is based on a specialized weighted moving least squares procedure. Time derivatives are formed by simple finite differences (implicit time integration). The point cloud moves according to the flow velocity, i.e. Lagrangian coordinates are used.

The resulting advantages are: (i) efficient handling of moving geometries, free surfaces, phase boundaries, or large deformations; (ii) low preprocessing costs (from CAD to simulation); (iii) easy integration of different material and custom (user-defined) models as well as refinement/coarsening criteria for the point cloud.

Details on the numerical scheme as well as the application of the method to selected physical problems can be found in [1–5, 9, 10]. Currently, FPM is further developed to an integrated tool called MESHFREE. It combines the above mentioned advantages with the robust and scalable solvers for large sparse linear systems of the SAMG library developed by Fraunhofer SCAI.

Due to their natural advantages, meshfree methods based on a Lagrangian formulation are frequently used to simulate physical problems that are characterized by Fluid Structure Interaction (FSI). In this contribution, we present different types of FSI in the context of GFDM/FPM. Thereby, we focus on the classical pressure–velocity coupling and the alternative velocity–pressure coupling to model the interaction of two point clouds or a point cloud with (flexible/rigid) structures (Sect. 3). The necessary approximation and integration strategies are described in Sect. 2. In conclusion, we discuss two industrial examples in Sect. 4: water–air coupling in Pelton turbines and flow in flexible tubes.

2 Basics of GFDM/FPM

In this section, we describe the approximation and integration strategies required to perform the necessary projection of quantities during Fluid Structure Interaction (FSI) applications in GFDM/FPM. Furthermore, the discretization of the computational domain by a cloud of numerical points is discussed.

2.1 Point Cloud

In contrast to other meshfree methods, the discretization in GFDM/FPM is based on a cloud of numerical points that act as carrier of physical information (pressure, velocity, temperature, etc.). This point cloud is generated and maintained in such a way that it compactly covers the whole computational domain Ω. Its density is given by a sufficiently smooth function in space and time which is called interaction radius:

$$h = h(\mathbf{x}, t). \tag{1}$$

In the first place, h is a user-given function defining the resolution of the point cloud and, thereby, the quality of the simulation results. Currently, we work on local adaptive refinement strategies for the point cloud, where h is constructed due to error estimates.

Based on the interaction radius h, the computational domain Ω is defined to be compactly covered by the point cloud if any ball with radius $r_{\text{hole}} \cdot h$ contains at least one point. For numerical reasons, r_{hole} is chosen such that for a point $\mathbf{x}_i$ of the point cloud $20 \ldots 50$ points are found within the interaction radius $h(\mathbf{x}_i, t)$ which directly determines the neighborhood of the point. $r_{\text{hole}} = 0.45$ is a common choice for practical applications.

Additionally, the distance between two points should not fall below a certain minimum: $r_{\text{small}} \cdot h$, where $r_{\text{small}} = 0.1 \ldots 0.2$.

The initial quality of the point cloud, which is ensured by the above criteria, has to be maintained during a simulation. This is due to the movement of the point cloud according to the flow velocity (Lagrangian framework) leading to accumulation or scattering of points. On the one hand, points have to be clustered if their distance is too small, i.e. the relevant points are merged according to specific rules (interior, free surface, boundary points). On the other hand, new points have to be filled if holes occur.

2.2 *Generalized Finite Difference Approximation*

The basic numerical idea of the FPM is a generalized Finite Difference (FD) approach. If the point cloud represented a perfectly regular point grid, then the approximation would work exactly like a classical FD method. As the point cloud is, however, non-regular in most cases, a generalized FD approach is used here.

2.2.1 Generalized Finite Difference Operators

We assume some point cloud being sufficiently dense, consisting of N points, whose positions are given by

$$\mathbf{x}_i = (x_i, y_i, z_i)^{\mathrm{T}}, \quad i = 1, \ldots, N. \tag{2}$$

Suppose furthermore that some arbitrary function f is represented by discrete function values at those discrete locations, i.e.

$$f_i = f(\mathbf{x}_i), \quad \mathbf{f} = (f_1, f_2, \ldots, f_N)^{\mathrm{T}}. \tag{3}$$

We call those vectors $\mathbf{c}_i^*$ the numerical differential operators which provide an approximation of some derivative (marked with *) in the sense

$$\partial_{\text{num}}^* f(\mathbf{x}_i) = \tilde{\partial}^* f(\mathbf{x}_i) = \tilde{\partial} f_i = \sum_{j=1}^{N} c_{ij}^* \cdot f_j = (\mathbf{c}_i^*)^{\mathrm{T}} \cdot \mathbf{f}. \tag{4}$$

The differential operator has to be established for every point $\mathbf{x}_i$ in the point cloud. In the scheme presented here, the most frequently used operators are:

$$\begin{aligned} \mathbf{c}_i^0 &= \text{ the numerical operator for function approximation,} \\ \mathbf{c}_i^x &= \text{ the numerical operator for the } x\text{-derivative,} \\ \mathbf{c}_i^y &= \text{ the numerical operator for the } y\text{-derivative,} \\ \mathbf{c}_i^z &= \text{ the numerical operator for the } z\text{-derivative,} \\ \mathbf{c}_i^\Delta &= \text{ the numerical operator for the Laplacian.} \end{aligned} \tag{5}$$

The aim is the development of operators that are independent of the underlying function values. Having operators that work generally for all given functions will save a lot of computation time.

We also introduce a weight function that switches on the particular neighbors close to some point $\mathbf{x}_i$ by

$$W_{ij} = w(r(\mathbf{x}_i, \mathbf{x}_j)) = w(r_{ij}), \tag{6}$$

where the distance function is given by

$$r(\mathbf{x}_i, \mathbf{x}_j) = r_{ij} = \frac{2 \cdot \|\mathbf{x}_i - \mathbf{x}_j\|}{h(\mathbf{x}_i) + h(\mathbf{x}_j)}. \tag{7}$$

We require $W(r \geq 1) = 0$ and sufficient smoothness of the weight function, such that the weight function "switches on" only neighbors in a close distance. The ansatz mostly used in our applications is

$$w(r) = \begin{cases} (1 - r^2)^\gamma, & \text{if } r < 1 \\ 0, & \text{otherwise} \end{cases}. \tag{8}$$

The discrete weights are grouped in the so-called weight matrix

$$\mathbf{W}_i = \begin{pmatrix} W_{i1} & & 0 \\ & W_{i2} & \\ & & \ddots & \\ 0 & & & W_{iN} \end{pmatrix} \tag{9}$$

having the discrete weights on its diagonal and zeros otherwise. Up to now, we have established the basis of the GFDM. The next section shows how to provide values to the differential operators.

2.2.2 Least Squares for Operator Generation

We are searching for the operator $\mathbf{c}_i^*$ for any particular differentiation task. For simplicity, we omit the star (*) and, instead of writing $\mathbf{c}_i^*$, we simply employ $\mathbf{c}_i$. The operator has to satisfy the least squares criterion

$$\frac{1}{2} \cdot \|\mathbf{W}_i^{-1} \cdot \mathbf{c}_i\|^2 = \frac{1}{2} \cdot \mathbf{c}_i^{\mathrm{T}} \cdot (\mathbf{W}_i^{-1})^{\mathrm{T}} \cdot \mathbf{W}_i^{-1} \cdot \mathbf{c}_i \stackrel{!}{=} \frac{1}{2} \cdot \mathbf{c}_i^{\mathrm{T}} \cdot \mathbf{W}_i^{-2} \cdot \mathbf{c}_i = \min \tag{10}$$

under the necessary consistency conditions

$$\mathbf{K}_i^{\mathrm{T}} \cdot \mathbf{c}_i = \mathbf{b}_i. \tag{11}$$

2.2.2.1 General Remark

The "·"-operator is the general matrix-times-matrix operator, where a matrix can also reduce to a vector or a scalar.

The matrix $\mathbf{K}_i$ represents column-wise, discrete test functions, for which the numerical operator shall give a distinct value. For example, the numerical operator for the x-derivative $\mathbf{c}_i^x$ shall deliver zero if operating on a constant function $\mathbf{k}_i^1 = (1, 1, \ldots, 1)^{\mathrm{T}}$ or a quadratic function $\mathbf{k}_i^3 = ((x_1-x_i)^2, (x_2-x_i)^2, \ldots, (x_N-x_i)^2)^{\mathrm{T}}$, but it shall deliver one if operating on the linear function $\mathbf{k}_i^2 = (x_1 - x_i, x_2 - x_i, \ldots, x_N - x_i)^{\mathrm{T}}$. In other words, we have the conditions

$$(\mathbf{k}_i^1)^{\mathrm{T}} \cdot \mathbf{c}_i^x = 0, \quad (\mathbf{k}_i^2)^{\mathrm{T}} \cdot \mathbf{c}_i^x = 1, \quad (\mathbf{k}_i^3)^{\mathrm{T}} \cdot \mathbf{c}_i^x = 0. \tag{12}$$

In general, the columns of matrix $\mathbf{K}_i$ represent M discrete test functions in the sense

$$\mathbf{K}_i \equiv (\mathbf{k}_i^1 \ \mathbf{k}_i^2 \ \mathbf{k}_i^3 \ \ldots \ \mathbf{k}_i^M). \tag{13}$$

Usually, a convenient choice for the $\mathbf{k}_i^p$, $p = 1, \ldots, M$, is to use the monomials up to a certain order, i.e. the numerical operator exactly reproduces these functions or their derivatives. In addition to that, in FPM, we have found that it is also necessary to determine the action of the operator on the delta function as well, i.e. we choose

$$\mathbf{k}_i^{M+1} = (\delta_{1i} \ \delta_{2i} \ \ldots \ \delta_{Ni})^{\mathrm{T}}. \tag{14}$$

Similarly to (12), the required results would be

$$(\mathbf{k}_i^{M+1})^{\mathrm{T}} \cdot \mathbf{c}_i^0 = C_0 < 1, \quad (\mathbf{k}_i^{M+1})^{\mathrm{T}} \cdot \mathbf{c}_i^x = 0, \quad (\mathbf{k}_i^{M+1})^{\mathrm{T}} \cdot \mathbf{c}_i^{xx} = C_{xx} = O\left(\frac{1}{h^2}\right). \tag{15}$$

The constants C_0 and C_{xx} have to be chosen adequately to gain optimal stability. In our simulations, $C_0 = 0.7$ and $C_{xx} = \frac{5}{h^2}$ are used.

The right hand side vector $\mathbf{b}$ consequently contains the corresponding values to be delivered by the operator if applied to the test functions.

The minimization problem (10) together with the consistency conditions (11) can be solved using Lagrangian multipliers in the sense

$$\mathbf{W}_i^{-2} \cdot \mathbf{c}_i - \sum_{k=1}^{M(+1)} \lambda_{ik} \cdot \mathbf{k}_i^k = 0 \tag{16}$$

which transforms to

$$\mathbf{c}_i = \sum_{k=1}^{M(+1)} \lambda_{ik} \cdot \mathbf{W}_i^2 \cdot \mathbf{k}_i^k. \tag{17}$$

This nicely shows that any differential operator is a linear combination of the weight-scaled test functions.

Multiplication from left with the test matrix $\mathbf{K}_i$ yields

$$\mathbf{K}_i^{\mathrm{T}} \cdot \mathbf{c}_i = \mathbf{K}_i^{\mathrm{T}} \cdot \mathbf{W}_i^{\mathrm{T}} \cdot \mathbf{W}_i \cdot \mathbf{K}_i \cdot \boldsymbol{\lambda}_i \tag{18}$$

which, by virtue of (11), provides

$$\mathbf{b}_i = \mathbf{K}_i^{\mathrm{T}} \cdot \mathbf{W}_i^{\mathrm{T}} \cdot \mathbf{W}_i \cdot \mathbf{K}_i \cdot \boldsymbol{\lambda}_i. \tag{19}$$

Equation (19) provides an $M \times M$ linear system to be solved for $\boldsymbol{\lambda}_i$ which then serves for providing the solution

$$\mathbf{c}_i = \mathbf{W}_i^2 \cdot \mathbf{K}_i \cdot \boldsymbol{\lambda}_i. \tag{20}$$

The most expensive part in terms of computation time is the generation of matrix $\mathbf{K}_i^{\mathrm{T}} \cdot \mathbf{W}_i^{\mathrm{T}} \cdot \mathbf{W}_i \cdot \mathbf{K}_i$ together with the solution of the linear system (19). There is one inherent risk which lies in the fact that we produce the square of the matrix $\mathbf{W}_i \cdot \mathbf{K}_i$. This matrix is usually well conditioned. Critical point cloud distortion might lead to singular or badly shaped matrices. One way to avoid this is to decrease the order of approximation (reduce the parameter M). On the other hand, better numerical conditioning of the local system (11) is achieved by a least norm approach instead of the Lagrange multiplier formulation presented above.

The procedure sketched in this section will fail if the point cloud locally deforms to a lower dimensional manifold, i.e. if (for a 3D computation) the points are nearly placed in a plane or on a line, etc. This happens for example if water squirts off a free surface. In this case, we strictly employ pseudo-inverse solutions to system (19) in order to keep the differential operators in good shape.

2.2.3 Solving PDEs on a GFDM-Basis

After having established numerical, meshfree operators for derivatives, we can provide strong form discretizations of partial differential equations. Our aim is to solve mass, momentum, and energy equations, the Navier–Stokes equations being a special form of them. For simplicity, let us have a look at a general convection diffusion equation of the form

$$\frac{D\Phi}{Dt} = \eta \cdot \Delta\Phi + q. \tag{21}$$

The numerical idea is to find a solution of the equation for every point in the point cloud. Using a time step size $\Delta t = t^{n+1} - t^n$, an explicit time integration scheme would read

$$\frac{\Phi_i^{n+1} - \Phi_i^n}{\Delta t} = \eta \cdot \sum_{j=1}^{M} c_{ij}^{\Delta} \cdot \Phi_j^n + q_i = \eta \cdot (\mathbf{c}_i^{\Delta})^{\mathrm{T}} \cdot \mathbf{\Phi}^n + q_i, \tag{22}$$

$$\mathbf{\Phi}^{n+1} = (\mathbf{I} + \Delta t \cdot \eta \cdot \mathbf{C}^{\Delta}) \cdot \mathbf{\Phi}^n + \Delta t \cdot \mathbf{q}. \tag{23}$$

The more convenient approach is the implicit time integration (25), as here, no stability restriction on the time step size has to be taken into account:

$$\mathbf{\Phi}^{n+1} = \mathbf{\Phi}^n + \Delta t \cdot \eta \cdot \mathbf{C}^{\Delta} \cdot \mathbf{\Phi}^{n+1} + \Delta t \cdot \mathbf{q}, \tag{24}$$

$$(\mathbf{I} - \Delta t \cdot \eta \cdot \mathbf{C}^{\Delta}) \cdot \mathbf{\Phi}^{n+1} = \mathbf{\Phi}^n + \Delta t \cdot \mathbf{q}. \tag{25}$$

In order to solve (25) for $\mathbf{\Phi}^{n+1}$, a sparse linear system of dimension $N \times N$ has to be solved. Each row of the linear systems (23) resp. (25) represents the collocated solution of the model PDE at one point in the cloud. If a point is located at the boundary, the row has to be replaced by appropriate boundary conditions of Dirichlet or Neumann type.

The matrix $(\mathbf{I} - \Delta t \cdot \eta \cdot \mathbf{C}^{\Delta})$ in (25) is invertible for all $\Delta t > 0$ if the matrix has M-character. The M-character itself depends on the local differential operators $\mathbf{c}_i^{\Delta}$, for which no a priori statement can be given. The use of the delta function as test function (see (14)) forces the M-character.

The implicit way of time integration can be applied to the Navier–Stokes equations or other models in a particularly natural way within the FPM-framework.

2.3 Integration of Simulation Results over Geometrical Entities

For the task of Fluid Structure Interaction (FSI) discussed in Sect. 3, we will require

- the mapping of function values, such as pressure or velocity, to the node points of an FE-mesh (if coupling to a mesh based solver),
- the mapping of function values to another point cloud of FPM (if the interacting structure is modeled also on a meshfree basis),
- the integration of function values over closed surfaces (such as rigid bodies) in order to compute integrated forces or moments.

This motivates the two tasks of mapping function values from the point cloud to (external) points and reduction of function values by integration.

2.3.1 Mapping Function Values

Suppose we need to map a function u, given by the discrete values $u_i, i = 1, \dots, N$, to the external point location $\mathbf{y}$. Our approach is using the approximation procedure described in Sect. 2.2:

$$\tilde{u}(\mathbf{y}) = \sum_{j=1}^{N} c_j^0(\mathbf{y}) \cdot u_j = (\mathbf{c}^0(\mathbf{y}))^{\mathrm{T}} \cdot \mathbf{u}. \tag{26}$$

Another notation used in this context is the mapping between two point clouds, the first one denoted by I and the second one by II, respectively. The mapping is marked in the following way:

- u_i^{I} is the discrete function value at point $\mathbf{x}_i^I$ of phase I and $\mathbf{u}^{\mathrm{I}}$ is the vector containing all discrete function values of this fluid phase.
- $u_i^{\mathrm{II}\to\mathrm{I}}$ is the discrete function value at point $\mathbf{x}_i^I$ of phase II mapped to phase I and $\mathbf{u}^{\mathrm{II}\to\mathrm{I}}$ is the vector containing all mapped discrete function values. The alternative notation, in order to improve readability, is $\mathbf{u}^{\mathrm{II}\to\mathrm{I}} \equiv \mathbf{u}^{\mathrm{II}}_{\to\mathrm{I}}$.

2.3.2 Reduction of Function Values by Integration

For some FSI-applications, such as coupling with rigid bodies, integrated forces and/or moments are necessary to be mapped from the point cloud to the structure.

The integrated force on a structure Ω is given by

$$F_\Omega = \oint_{\partial\Omega} (p \cdot \mathbf{n} - \mathbf{S} \cdot \mathbf{n}) \, d(\partial\Omega) \tag{27}$$

which is approximated by the sum over all boundary points touching the surface of Ω:

$$F_\Omega \approx \sum_{i \in \partial\Omega} (p_i \cdot \mathbf{n}_i - \mathbf{S}_i \cdot \mathbf{n}_i) \cdot A_i. \tag{28}$$

This requires the knowledge of the area A_i represented by a boundary point. We produce the boundary area entity of a boundary point by a local Voronoi decomposition around this point and choose the area of the Voronoi face $\tilde{A}_i$ as a first guess. This area is corrected in a least squares sense $\tilde{A}_i \rightarrow A_i$ such that the following consistency constraints are satisfied:

$$\begin{aligned} \sum_{\mathbf{x}_i \in \partial\Omega} A_i &= A(\partial\Omega), \\ \sum_{\mathbf{x}_i \in \partial\Omega} (\mathbf{e}_x^T \cdot \mathbf{n}_i) A_i = \sum_{\mathbf{x}_i \in \partial\Omega} (\mathbf{e}_y^T \cdot \mathbf{n}_i) A_i &= \sum_{\mathbf{x}_i \in \partial\Omega} (\mathbf{e}_z^T \cdot \mathbf{n}_i) A_i = 0, \\ \sum_{\mathbf{x}_i \in \partial\Omega} ((x_i, 0, 0)^T \cdot \mathbf{n}_i) A_i = \sum_{\mathbf{x}_i \in \partial\Omega} ((0, y_i, 0)^T \cdot \mathbf{n}_i) A_i &= \sum_{\mathbf{x}_i \in \partial\Omega} ((0, 0, z_i)^T \cdot \mathbf{n}_i) A_i = V(\Omega), \end{aligned} \tag{29}$$

where $\mathbf{e}_x, \mathbf{e}_y, \mathbf{e}_z$ are the unit vectors in x-,y-,z-direction, respectively.

3 Fluid Structure Interaction (FSI) in GFDM/FPM

The interaction of a fluid/continuum with a structure can be modeled in two different ways. Depending on the inertial forces of the structure, either a pressure–velocity coupling or a velocity–pressure coupling can be used.

3.1 Pressure–Velocity Coupling

In the classical ansatz, the fluid/continuum provides its pressure and tension to the structure. Then, the structure model is solved and provides back its kinematic or dynamic behavior (i.e. velocity and position) as a boundary condition to the fluid/continuum. According to the type of interaction zone, a surface–surface coupling or a volume–volume coupling has to be chosen.

3.1.1 Surface–Surface Coupling

If the fluid/continuum and the structure (or second fluid/continuum) interact by a clearly defined common surface, the surface–surface coupling should be used. An air bubble within a bath of water, a fluttering sheet of paper in the wind, or a drilling tool entering a metal part could be representative examples of this type of phenomenon. Some of our industrially relevant examples are described in the following.

3.1.1.1 Explicit Approach

We consider classical airbag inflation simulations in car crash scenarios. Here, GFDM/FPM models the gas dynamics inside of the airbag and produces pressure p and turbulent tension τ_W at the airbag membrane, and some FE-solver integrates the dynamic behavior of the airbag membrane based on the computed wall pressure/tension values. Let us name the FPM-phase with the index I and the structure phase by the index S. Then, the explicit scheme reads

$$\begin{cases} (p^{\mathrm{I}}, \mathbf{v}^{\mathrm{I}}, \tau_W^{\mathrm{I}})(t^{n+1}) = \text{FPMsolver}(\Delta t;\, p^{\mathrm{I}}(t^n), \mathbf{v}^{\mathrm{I}}(t^n)), \\ \mathbf{v}^{\mathrm{I}}(t^{n+1})\big|_{\text{interface}} = \mathbf{v}^{\mathrm{S}}_{\to \mathrm{I}}(t^n), \end{cases}$$
$$\mathbf{v}^{\mathrm{S}}(t^{n+1}) = \text{FEsolver}(\Delta t;\, \mathbf{v}^{\mathrm{S}}(t^n), p^{\mathrm{I}}_{\to \mathrm{S}}(t^n), \tau_W{}^{\mathrm{I}}_{\to \mathrm{S}}(t^n)), \tag{30}$$

where $\Delta t = t^{n+1} - t^n$ is, us usual, the time step size. FPMsolver and FEsolver contain the classical boundary conditions at the non-interface boundaries. The upper two equations are solved first, then the lower one.

3.1.1.2 Implicit Approach

We model implicit coupling only within the framework of GFDM/FPM. Coupling FPM with another code would mean to produce joint systems of equations, an almost impossible task when coupling with commercial FE-codes.

The implicit coupling scheme consists of solving two phases separately on the classical FPM-basis, with additional requirements at the surface–surface interface. Let as name the two phases with the indices I and II, respectively. The implicit scheme can be written as

$$\begin{cases} (p^{\mathrm{I}}, \mathbf{v}^{\mathrm{I}})(t^{n+1}) = \mathrm{FPMsolver}(\Delta t;\, p^{\mathrm{I}}(t^n), \mathbf{v}^{\mathrm{I}}(t^n)), \\ (p^{\mathrm{II}}, \mathbf{v}^{\mathrm{II}})(t^{n+1}) = \mathrm{FPMsolver}(\Delta t;\, p^{\mathrm{II}}(t^n), \mathbf{v}^{\mathrm{II}}(t^n)), \\ p^{\mathrm{I}} = p^{\mathrm{II}}_{\to\mathrm{I}}\big|_{\mathrm{interface}}, \\ \dfrac{\tilde{\partial} p^{\mathrm{II}}}{\tilde{\partial} n} = -\dfrac{\tilde{\partial} p^{\mathrm{I}}_{\to\mathrm{II}}}{\tilde{\partial} n}\Big|_{\mathrm{interface}}, \\ \dfrac{\tilde{\partial} \mathbf{v}^{\mathrm{I}}}{\tilde{\partial} n} = -\dfrac{\tilde{\partial} \mathbf{v}^{\mathrm{II}}_{\to\mathrm{I}}}{\tilde{\partial} n}\Big|_{\mathrm{interface}}, \\ \mathbf{v}^{\mathrm{II}} = \mathbf{v}^{\mathrm{I}}_{\to\mathrm{II}}\big|_{\mathrm{interface}}. \end{cases} \tag{31}$$

Our typical example here is fuel–air coupling in tank-filling and tank-sloshing scenarios. Of course, it can be applied only if the phase boundary between air and fuel stays regular and does not spray out.

3.1.2 Volume–Volume Coupling

In case of two phases (fluids/continua) that are characterized by complex free surfaces impeding the projection of the necessary quantities between the phases, a volume–volume coupling can be used as an alternative to the above described surface–surface coupling. Thereby, both phases interact by a volume overlap.

The basic idea of the volume–volume coupling is that the two phases as well as the corresponding point clouds are separate and only exchange necessary information by approximation/projection from one phase to the other phase. Thereby, modeling the actual contact between the two phases as in the surface–surface coupling is replaced by the following procedure: Only one phase models the complex free surfaces (phase II), while the other is overlapping the more complex one (phase I). For the coupling, the Brinkman model is used (see [8] and the references therein).

3.1.2.1 Phase I

For the overlapping phase, the standard formulation of the conservation of momentum in GFDM/FPM is extended to

$$\frac{d\mathbf{v}^{\mathrm{I}}}{dt} + \frac{1}{\rho} \cdot \nabla p^{\mathrm{I}} = \frac{1}{\rho} \cdot (\nabla^{\mathrm{T}} \mathbf{S}^{\mathrm{I}})^{\mathrm{T}} + \mathbf{g} - \eta \cdot (\mathbf{v}^{\mathrm{I}} - \mathbf{v}^{\mathrm{II}}_{\to\mathrm{I}}), \tag{32}$$

where $\nabla^{\mathrm{T}} \mathbf{S}^{\mathrm{I}}$ is the divergence of the stress tensor in phase I and $\mathbf{g}$ is the vector of body forces. The drag coefficient η and the reference velocity $\mathbf{v}_{\mathrm{ref}}$ are determined by phase II. In other words, the overlapping phase flows through a porous medium whose properties are defined by the other phase. The computational domain of phase I as well as phase II below is defined by the specific application.

In the following, the influence of the local presence of phase II on the choice of η is discussed:

- If there are no points of phase II in the local neighborhood of a point of phase I, the flow is unobstructed by the other phase at this location and the drag coefficient for this point is zero, i.e. $\eta = 0\,\mathrm{s}^{-1}$.
- If the local neighborhood in phase II of a point in phase I is non-empty, the drag coefficient depends on the type of neighborhood that is found.
  - In case of a bulk-like neighborhood, the value of η is chosen appropriately large at this location to model an impermeable medium, e.g. $\eta = 10^4\,\mathrm{s}^{-1}$.
  - In case of a spray-like free surface neighborhood, the drag is determined based on the assumption of spheric droplets. For water, this leads to

$$\eta = \frac{3}{4} \cdot \alpha^{\mathrm{II}}_{\to \mathrm{I}} \cdot C_{\mathrm{D}} \cdot \frac{\|\mathbf{v}^{\mathrm{I}} - \mathbf{v}^{\mathrm{II}}_{\to \mathrm{I}}\|}{d^{\mathrm{II}}_{\to \mathrm{I}}}, \tag{33}$$

where $\alpha^{\mathrm{II}}_{\to \mathrm{I}}$ is the volume fraction and $d^{\mathrm{II}}_{\to \mathrm{I}}$ is the diameter of the droplets in phase II mapped to phase I. Furthermore, we have $C_{\mathrm{D}} = \max\left(\frac{24}{\mathrm{Re}}, 1\right)$ with relative Reynolds number $\mathrm{Re} = \frac{\rho^{\mathrm{I}} \cdot \|\mathbf{v}^{\mathrm{I}} - \mathbf{v}^{\mathrm{II}}_{\to \mathrm{I}}\| \cdot d^{\mathrm{II}}_{\to \mathrm{I}}}{\eta^{\mathrm{I}}}$.

3.1.2.2 Phase II

The physical model for phase II corresponds to the classical conservation equations in GFDM/FPM. The momentum equation is given by

$$\frac{d\mathbf{v}^{\mathrm{II}}}{dt} + \frac{1}{\rho} \cdot \nabla p^{\mathrm{II}} = \frac{1}{\rho} \cdot (\nabla^{\mathrm{T}} \mathbf{S}^{\mathrm{II}})^{\mathrm{T}} + \mathbf{g}^{\mathrm{II}}. \tag{34}$$

Similar to the procedure described above, the influence of phase I is integrated depending on the local configuration of the point cloud in phase II.

- In case of a bulk-like free surface configuration, i.e. phase II is like an impermeable medium for phase I, the stagnation pressure of phase I acting on the local free surface of phase II is used as Dirichlet boundary condition:

$$p^{\mathrm{II}}\Big|_{\mathrm{FS}} = \frac{1}{2} \cdot \rho^{\mathrm{I}}_{\to \mathrm{II}} \cdot \left((\mathbf{v}^{\mathrm{I}}_{\to \mathrm{II}} - \mathbf{v}^{\mathrm{II}}) \cdot \mathbf{n}^{\mathrm{II}} \right)^2. \tag{35}$$

- In case of a spray-like free surface configuration, additional forces act on the single droplets which can be modeled by extending the body forces term according to the procedure leading to (33):

$$\mathbf{g}^{\mathrm{II}} = \mathbf{g} - \frac{3}{4} \cdot \frac{\rho^{\mathrm{I}}_{\to \mathrm{II}}}{\rho^{\mathrm{II}}} \cdot C_{\mathrm{D}} \cdot \frac{\|\mathbf{v}^{\mathrm{I}}_{\to \mathrm{II}} - \mathbf{v}^{\mathrm{II}}\|}{d^{\mathrm{II}}} \cdot (\mathbf{v}^{\mathrm{I}}_{\to \mathrm{II}} - \mathbf{v}^{\mathrm{II}}). \tag{36}$$

The coupling terms in (35) and (36) are based on projected values of phase I.

Industrially relevant applications for this type of coupling are manifold. Two examples from automotive industry are given below.

Explicit Approach Water entering the air-intake of a car engine at deep-water crossing of a vehicle.

Implicit Approach Spray/droplet clouds in the airflow around a vehicle.

3.2 Velocity–Pressure Coupling

If the structure (such as a membrane) tends to have zero inertial forces, the numerical time step size dramatically drops and the classical pressure–velocity coupling described in Sect. 3.1 becomes problematic. In this case, the alternative velocity–pressure coupling is more stable. It can be considered as an enhanced surface tension formulation: the fluid/continuum provides the movement of the structure, while the state of inner stresses within the structure results in a boundary condition for pressure to the fluid/continuum.

Explicit Approach
From the old time step, we inquire the resulting pressure of the structure and apply it as a boundary condition to the pressure of the new time cycle. Let us name the FPM phase with the index I, and the structure phase with the index S. Then, the explicit scheme reads

$$\begin{cases} (\mathbf{x}^{\mathrm{I}}, p^{\mathrm{I}}, \mathbf{v}^{\mathrm{I}})(t^{n+1}) = \mathrm{FPMsolver}(\Delta t; \mathbf{x}^{\mathrm{I}}(t^n), p^{\mathrm{I}}(t^n), \mathbf{v}^{\mathrm{I}}(t^n)), \\ p^{\mathrm{I}}(t^{n+1})\big|_{\mathrm{interface}} = p^{\mathrm{S}}_{\to \mathrm{I}}(t^n), \end{cases}$$
$$p^{\mathrm{S}}(t^{n+1}) = \mathrm{FEsolver}(\Delta t; \mathbf{x}^{\mathrm{I}}_{\to \mathrm{S}}(t^{n+1}), \mathbf{v}^{\mathrm{I}}_{\to \mathrm{S}}(t^{n+1})). \tag{37}$$

As the scheme of coupling is explicit, there is a time step constraint that depends on the sensitivity of the structure (pressure changes induced) based on small changes of the local positions along the interface. The time step constraint has the form

$$\Delta t < \sqrt{\frac{1}{\sigma} \cdot \rho \cdot h^3}, \tag{38}$$

where σ is the representative inner tension of the structure.

This approach can be industrially used for flows in textile liners, blood vessels, or for blow-deformation processes in container production.

Implicit Approach
Equation (38) shows that the time step size becomes very small if the inner tension of the structure becomes large and the numerical resolution becomes fine. In order to gain larger time step sizes, an implicit approach would be required. One could

iterate the scheme (37) and hope to find a fixed point. However, if there exists a linearized approximation of the pressure change with respect to the position change at the interface $\frac{\partial p^{\mathrm{S}}}{\partial \mathbf{x}}$, then an implicit formulation is possible:

$$\begin{cases} (\mathbf{x}^{\mathrm{I}}, p^{\mathrm{I}}, \mathbf{v}^{\mathrm{I}})(t^{n+1}) = \mathrm{FPMsolver}(\Delta t; \mathbf{x}^{\mathrm{I}}(t^n), p^{\mathrm{I}}(t^n), \mathbf{v}^{\mathrm{I}}(t^n)), \\ p^{\mathrm{I}}(t^{n+1})\big|_{\mathrm{interface}} = p^{\mathrm{S}}_{\to \mathrm{I}}(t^n) + \Delta t \cdot \frac{\partial p^{\mathrm{S}}}{\partial \mathbf{x}} \cdot \mathbf{v}^{\mathrm{I}}(t^{n+1}), \\ p^{\mathrm{S}}(t^{n+1}) = \mathrm{FEsolver}(\Delta t; \mathbf{x}^{\mathrm{I}}_{\to \mathrm{S}}(t^{n+1}), \mathbf{v}^{\mathrm{I}}_{\to \mathrm{S}}(t^{n+1})). \end{cases} \tag{39}$$

The second line in the scheme can be linearly incorporated into the FPM-solver in order to form an enhanced linear system.

Potential applications of the scheme are rain/water drops of small Weber-numbers inside of complex geometries' water drain simulations or membrane pumps in medical applications.

4 Industrial Applications

In this section we focus on two specific applications in the context of FSI: Pelton turbine applications and flow in flexible tubes. We discuss the main aims of the respective simulations as well as the occurring challenges in terms of FSI.

Concerning the least squares operators, we use monomials up to second order as well as the extension by the delta function in (11).

4.1 Water–Air Coupling in Pelton Turbines

Pelton turbines are impulse-type water turbines characterized by a free jet of water impacting with the runner of the turbine. The main aim of simulations in this field of application is to assist the turbine development during modernization projects of existing hydropower plants. In this process, replacements of the runner and parts of the housing have to meet contractual values. Different demands have to be met by the simulations:

- modeling of single water droplets
- analysis of torques on specific parts of the geometry (runner, deflector, etc.)
- filling of a nozzle including the generation of the water jet
- coupling of water phase and air phase including an intelligent time stepping criterion
- coupling of water phase and granular phase including an abrasion model

In the following, we solely focus on the coupling of water and air phase. This problem can be modeled by the pressure–velocity coupling described in Sect. 3.1.

Due to the complex free surfaces (water jet, water sheets, and single water droplets), the volume–volume coupling approach is chosen.

The point clouds for the water and the air phase are generated and maintained independently from each other. At any time, the seeding of points is based on a discrete hole search according to the specified point cloud parameters (see Sect. 2.1). A detailed description of the discrete hole search can be found in [6, 7].

- *Water phase:* Points on the circular inflow boundary are seeded initially starting from the edge until the inside of the surface is filled. Based on these boundary points, further points are seeded in order to represent a small initial water jet. This is realized by duplicating and moving selected points in the normal direction of the inflow boundary (free surface points) and, subsequently, filling the interior (interior points).
  In the course of the simulation, the points move according to the solution of the Navier–Stokes equations and the defined boundary conditions. Additionally, points are injected at the inflow boundary with the corresponding inflow velocity based on the interaction radius h. The point cloud is maintained by regularly checking for holes and accumulations.
- *Air phase:* Similarly, boundary points are first seeded on the housing, the nozzles (including the inflow), the runner, and the built-ins of the turbine geometry. Subsequently, the interior is filled with points.
  In this case, the points move only due to the solution of the Navier–Stokes equations. There is no injection at the inflow which is deemed as a standard wall boundary.

For both phases, points leaving the turbine at the lower or the right-hand boundary are deleted.

Decoupling the water and the air phase enables different solution strategies for the two phases. For the water phase, conservation of mass and momentum can be solved for the pressure and velocity in a coupled ansatz using a GFDM/FPM-specific penalty formulation. In contrast to that, a classical segregated ansatz based on Chorin's projection idea is used to solve for pressure and velocity for the air phase. See [3] for details on these solution strategies.

Furthermore, different point cloud resolutions h can be used—high density of points in the water phase and lower density of points in the air phase. Since the phase-specific time step sizes mostly depend on the chosen interaction radius h for this kind of application, they may differ up to one order of magnitude. Here, the time step size in the water phase is always smaller than the one in the air phase. Thus, an intelligent time stepping criterion including a synchronization process between the phases is needed. The global time step size Δt_{global} is defined by

$$\Delta t_{\text{global}} = \min\left(\Delta t_{\text{water}}, \Delta t_{\text{air}}\right) = \Delta t_{\text{water}}, \tag{40}$$

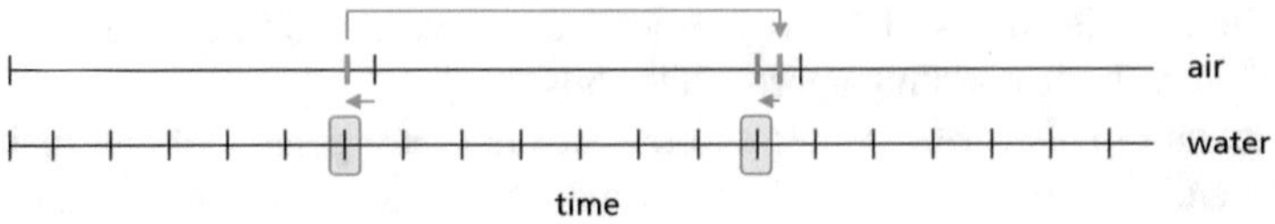

Fig. 1 Schematic illustration of the synchronization process of the phase-specific time step sizes

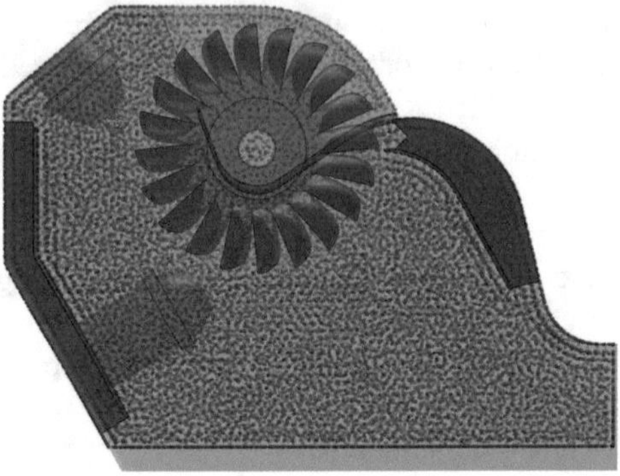

Fig. 2 Horizontal Pelton turbine with one active inflow—simulation setup

where the phase-specific time step size for the air phase is synchronized to the one of the water phase in case of $t^n_{\text{water}} + \gamma \cdot \Delta t_{\text{water}} > t^n_{\text{air}} + \Delta t_{\text{air,potential}}$ ($\gamma > 1$; here, $\gamma = 2$):

$$\Delta t_{\text{air}} = t^n_{\text{water}} + \Delta t_{\text{water}} - t^n_{\text{air}} \tag{41}$$

with n denoting the current phase-specific time step level.[1] In other words, a new solution for the air phase is only triggered if the potential phase-specific time step size of the air phase would lead to a phase-specific time that is larger but too close to the one of the water phase. If no new solution of the air phase is triggered, the water phase uses the air phase values of the previous air time step. This procedure is illustrated in Fig. 1.

As example, we consider a horizontal Pelton turbine with two inflows of which only one is active, see Fig. 2. The inflow velocity of the water phase is 43.4 m/s. The initial velocity of the air phase is zero. During the simulation, the velocity of the air phase increases due to the rotation of the runner and the coupling with the water phase to approximately 30–45 m/s in the vicinity of the runner and in the overlapping region with bulk-like water phase configurations, respectively. Choosing constant interaction radii $h_{\text{water}} = 0.006$ m and $h_{\text{air}} = 0.06$ m, the above discussed time stepping procedure leads to solving for the air phase only every 5–10 water phase time steps.

[1]For reasons of readability, we write n instead of n_{water} and n_{air}, respectively.

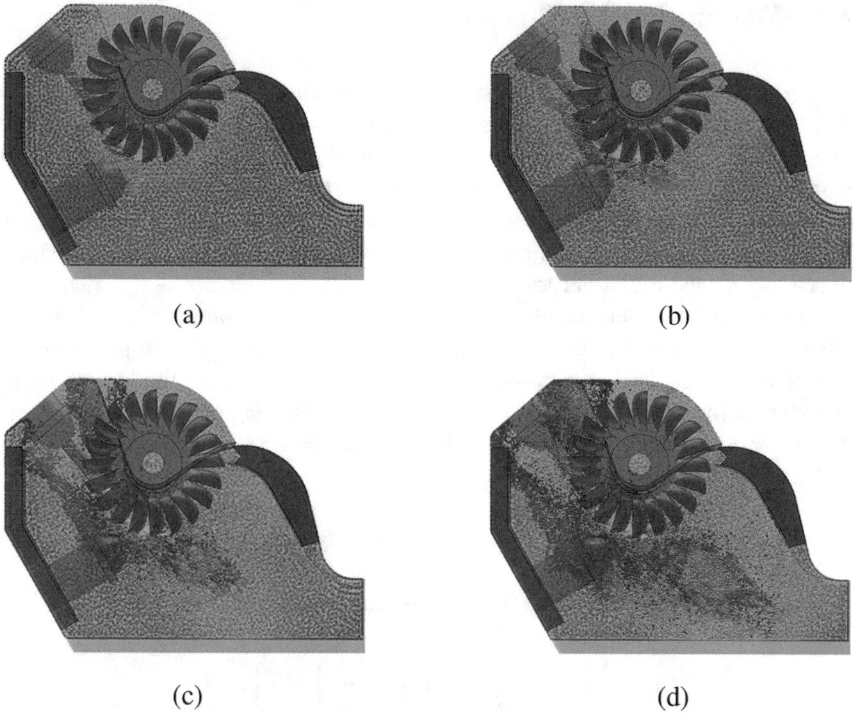

Fig. 3 Horizontal Pelton turbine with one active inflow—simulation evolution (a) $t = 0.015$ s; (b) $t = 0.03$ s; (c) $t = 0.045$ s; (d) $t = 0.06$ s

The temporal evolution of the simulation is illustrated in Fig. 3 (the points are colored according to their current velocity). The two-way coupling of water and air phase leads to the following effects: In the overlapping region with the bulk-like water phase, the air phase is pulled along with the flow velocity of the water. In the overlapping regions with sheet-like and spray-like water phase, the influence on the air phase is a lot smaller. However, the behavior of the water phase is significantly changed in these regions. Especially for the secondary flow moving from the runner to the top of the housing, the influence of the air phase on the water phase is essential to realistically evaluate the turbine based on the simulation. If this secondary flow or part of it is pushed back onto the runner, the efficiency of the turbine is reduced.

This simulation is part of a feasibility study. Model validation based on a comparison with video recording of a laboratory test is in progress.

4.2 Flow in Flexible Structures

This section demonstrates the capability of FPM/GFDM regarding flows within flexible structures and hulls, solved by the velocity–pressure coupling introduced in Sect. 3.2. Potential applications are flow in blood vessels, inflation of textile structures by foams such as car seats, light weight construction elements etc.

For the present showcase, we use a simple, cylindrical tube of radius r_0 that is bounded by an elastic membrane. This membrane changes its inner tension depending on the radial deformation, i.e. $\boldsymbol{\tau} = (\tau_a, \tau_b)^{\mathrm{T}} = \boldsymbol{\tau}(r - r_0)$. If $\mathbf{n}$ is the normal direction of the membrane, $\mathbf{a}$ and $\mathbf{b}$ are the two mutually perpendicular tangential directions. τ_a and τ_b are the tension components with respect to these directions. Let us further suppose that we are able to compute the instantaneous curvature components κ_a and κ_b of the membrane in the tangential directions. Then, the pressure boundary condition, that is required for the explicit coupling (37) as well as for the implicit scheme (39), is

$$p^S = \kappa_a \cdot \tau_a + \kappa_b \cdot \tau_b. \tag{42}$$

The inflow conditions are given as a pulsating velocity in the sense

$$v_{\text{inflow}} = \frac{v_0}{2} \cdot \left(\cos\left(\frac{2\pi t}{T_{\text{cycle}}} \right) + 1 \right), \tag{43}$$

where T_{cycle} is the cycle time of the process and v_0 the representative speed.

The generation and maintenance of the point cloud is performed analogously to the one described in Sect. 4.1. First, points are seeded on the inflow and outflow boundaries (boundary points) as well as the elastic membrane (free surface points). Subsequently, the inside is filled from the boundaries. In the course of the simulation, points are moved according to the solution of the Navier–Stokes equations. Furthermore, points are injected at the inflow boundary.

At the inflow boundary (left in each frame in Fig. 4), the points are colored according to the current inflow speed once they are injected. Red means an injection speed of v_0, blue means an injection speed of zero. After the initial coloring, the color is kept fixed throughout the simulation. Thus, the flow in the flexible pipe (and the history of the material transport) can be intuitively illustrated. The difference between bulk transport inside the pipe and the sound waves of the pipe membrane are obvious.

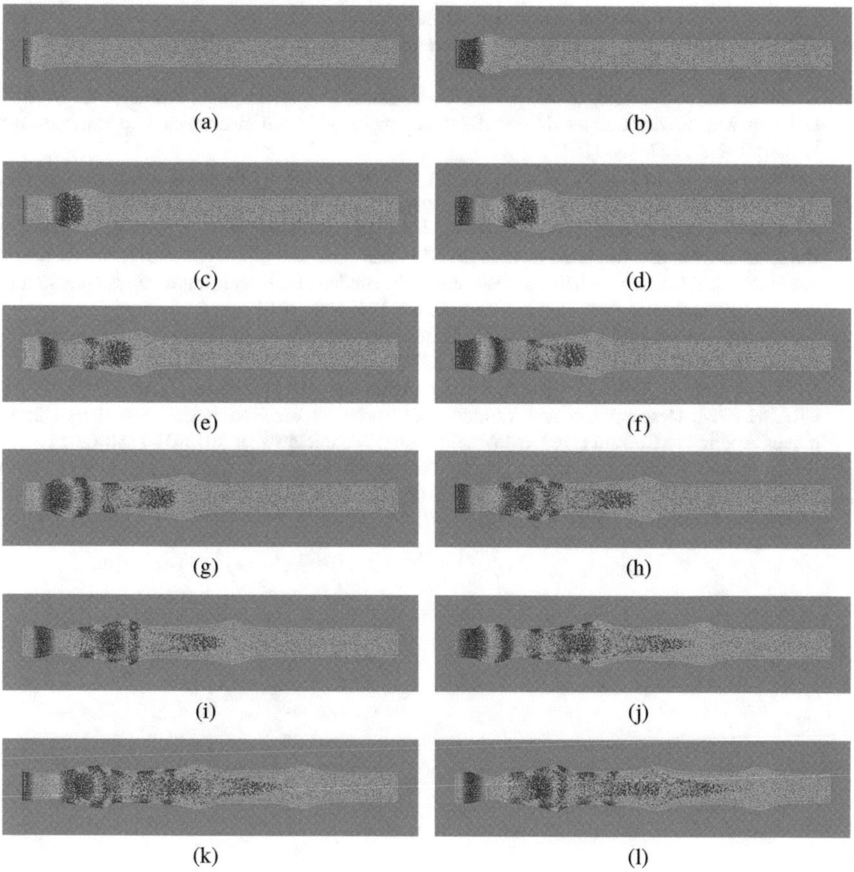

Fig. 4 Transport of colorized material inside a flexible pipe—simulation evolution (**a**) $t = 0$; (**b**) $t = 0.25 \times T_{\text{cycle}}$; (**c**) $t = 0.50 \times T_{\text{cycle}}$; (**d**) $t = 0.75 \times T_{\text{cycle}}$; (**e**) $t = 1.00 \times T_{\text{cycle}}$; (**f**) $t = 1.25 \times T_{\text{cycle}}$; (**g**) $t = 1.50 \times T_{\text{cycle}}$; (**h**) $t = 1.75 \times T_{\text{cycle}}$; (**i**) $t = 2.00 \times T_{\text{cycle}}$; (**j**) $t = 2.25 \times T_{\text{cycle}}$; (**k**) $t = 2.50 \times T_{\text{cycle}}$; (**l**) $t = 2.75 \times T_{\text{cycle}}$

References

1. A. Jefferies, J. Kuhnert, L. Aschenbrenner, U. Giffhorn, Finite pointset method for the simulation of a vehicle travelling through a body of water, in *Meshfree Methods for Partial Differential Equations VII*, ed. by M. Griebel, M. Schweitzer. Lecture Notes in Computational Science and Engineering, vol. 100 (Springer, Cham, 2015)
2. J. Kuhnert, Finite pointset method (FPM): meshfree flow solver with applications to elasto-plastic material laws, in *Proceedings First International Conference on Particle-Based Methods, PARTICLES 2009*, ed. by E. Oñate, D.R.J. Owen (CIMNE, Barcelona, 2009), pp. 423–426
3. J. Kuhnert, Meshfree numerical schemes for time dependent problems in fluid and continuum mechanics, in *Advances in PDE Modeling and Computation*, ed. by S. Sundar (Ane Books, New Delhi, 2014)

4. I. Michel, S.M.I. Bathaeian, J. Kuhnert, D. Kolymbas, C.-H. Chen, I. Polymerou, C. Vrettos, A. Becker, Meshfree generalized finite difference methods in soil mechanics. Part II: numerical results. Int. J. Geosci. **8**(2), 191–217 (2017)
5. I. Ostermann, J. Kuhnert, D. Kolymbas, C.-H. Chen, I. Polymerou, V. Smilauer, C. Vrettos, D. Chen, Meshfree generalized finite difference methods in soil mechanics. Part I: theory. Int. J. Geosci. **4**(2), 167–184 (2013)
6. S. Schröder, I. Michel, T. Seidel, C.M. König, STRING 3: full 3D visualization of groundwater flow, in *Proceedings of IAMG 2015—17th Annual Conference of the International Association for Mathematical Geosciences*, ed. by H. Schaeben, R. Tolosana Delgado, K.G. van den Boogaart, R. van den Boogaart (2015), pp. 813–822
7. T. Seidel, C. König, M. Schäfer, I. Ostermann, T. Biedert, D. Hietel, Intuitive visualization of transient groundwater flow. Comput. Geosci. **67**, 173–179 (2014)
8. T. Seifarth, Numerische Algorithmen für Gitterfreie Methoden zur Lösung von Transportproblemen, PhD thesis, Universität Kassel, 2018
9. A. Tramecon, J. Kuhnert, Simulation of Advanced Folded Airbags with VPS PAM-CRASH/FPM: Development and Validation of Turbulent Flow Numerical Simulation Techniques Applied to Curtain Bag Deployments, SAE Technical Paper 2013-01-1158 (2013)
10. E. Uhlmann, R. Gerstenberger, J. Kuhnert, Cutting simulation with the meshfree finite pointset method. Proc. CIRP **8**(Suppl. C), 391–396 (2013). 14th CIRP Conference on Modeling of Machining Operations (CIRP CMMO)

Parallel Detection of Subsystems in Linear Systems Arising in the MESHFREE Finite Pointset Method

Fabian Nick, Hans-Joachim Plum, and Jörg Kuhnert

Abstract The Finite Pointset Method (FPM) is a meshfree method for simulations in the field of fluid dynamics and continuum mechanics (Tiwari and Kuhnert, Finite pointset method based on the projection method for simulations of the incompressible Navier–Stokes equations. Springer, Berlin, 2003). The key idea in FPM is to discretize the necessary differential operators by using stencils generated by a least squares approach on a pointcloud that is moving in every time step.

Applying Algebraic Multigrid Methods (AMG) to the linear systems arising in FPM comes with various challenges, see our previous work Metsch et al. (Comput Vis Sci, reviewed) and Nick et al. (Linear solvers for the finite pointset method. In: Schäfer, M., Behr, M., Mehl, M., Wohlmuth, B. (eds.) Recent advances in computational engineering. Springer, Cham, 2018). In Nick et al. (Linear solvers for the finite pointset method. In: Schäfer, M., Behr, M., Mehl, M., Wohlmuth, B. (eds.) Recent advances in computational engineering. Springer, Cham, 2018) we limited ourselves to essentially irreducible matrices, saying that if a matrix arising from FPM is not essentially irreducible, we can employ a parallel algorithm in order to detect those subsystems that are essentially irreducible. This paper introduces the algorithm that we use in order to detect independent parts of the FPM pointcloud, which we call *components*. The algorithm that we propose has a theoretical complexity of $O(|V|)$ in the average case, where $|V|$ is the number of points in the pointcloud. Our experiments with a real world model however show that in practice the complexity is much better.

The experiments also show that in order to guarantee a stable convergence of the arising linear systems, detecting components is essential, as singular components can occur in certain situations.

F. Nick (✉) · H.-J. Plum
Fraunhofer SCAI, Sankt Augustin, Germany
e-mail: fabian.nick@scai.fraunhofer.de; hans-joachim.plum@scai.fraunhofer.de

J. Kuhnert
Fraunhofer ITWM, Kaiserslautern, Germany
e-mail: joerg.kuhnert@itwm.fraunhofer.de

M. Griebel, M. A. Schweitzer (eds.), *Meshfree Methods for Partial Differential Equations IX*, Lecture Notes in Computational Science and Engineering 129,
https://doi.org/10.1007/978-3-030-15119-5_6

Finally, we give an outlook on how our handling of the components could be improved in the future.

1 Introduction

Like in other discretization methods, the linear solver in FPM can become the main bottleneck for the performance of the overall method, if the problem either gets very large or the linear system is close to a saddle point problem. The latter case can occur for simulations with FPM using the coupled approach, see [20] for an introduction to the method and [11] for an analysis of those systems. But even when using the segregated approach the linear solver can become a bottleneck if the systems get larger. This is mainly because classical one level iterative methods do not scale linearly in the problem size. The BiCGStab(2) method [13] that serves as a baseline in our studies is no different in this respect.

AMG is known to be a very fast solver for sparse linear systems arising from elliptic PDEs and scales linearly in the problem size. In the setup phase, a hierarchy of linear systems is built with information from the matrix itself only and this hierarchy is used in the solving phase to solve the linear system.

As outlined in [12], applying AMG to the linear systems arising in FPM comes with some challenges, one of which is that the FPM discretization might lead to unwanted, independent subsystems (components) in the matrix that can be singular in the worst case. Section 2 will introduce the basics and notation from graph theory that we need to define what a component is. We will then touch on some related work before explaining why we observe components in some simulations conducted with FPM. The algorithm we give in Sect. 2.4 is tailored to the structure of the graphs we are considering in this context and its complexity is analyzed in Sect. 2.5. Section 2.6 briefly states how we exploit the information gained from the components detection in our linear solver before we give some numerical results from a real world example in Sect. 2.7. Possible remedies for the main drawback that becomes apparent in this section are discussed in Sect. 3.

2 Components

The dynamic character of a moving pointcloud implies a varying neighborhood structure in time. The local changes in the neighborhoods can lead to *groups* (or *components*, see Sect. 2.1) of points that are *independent* from the main pointcloud. This means that there are no connections to or from those independent groups of points. There are various reasons for this phenomenon, see Sect. 2.3. In the end, in all cases we end up with a linear system that can be decomposed into multiple subsystems that are also independent from each other. We can, or in some cases have to, take advantage of that property when solving the linear system. To this end, we use the algorithm described in Sect. 2.4 which finds such independent subsystems, even if they are distributed across multiple processes.

2.1 Graph Theory Basics and Notation

In order to detect independent groups of points within the pointcloud $\mathcal{P}$, we examine the graph $\bar{G}(A)$ associated to the matrix A that represents the linear system $Ap = g$ that is used to solve for one of the pressure fields needed in FPM [20]. A similar analysis can be carried out for the systems corresponding to the velocity equations in FPM. We make use of the special structure of those systems in our algorithm but will not go into detail here in this work.

Definition 1 A *graph* is a pair $G = (V, E)$ of two sets V and E, where the elements of V are called *vertices* and the elements of E are called *edges*, where

- $E \subset \{(v, w) : i, v \in V\}$ for *directed* graphs and
- $E \subset \{\{v, v\} : v, w \in V\}$ for *undirected* graphs

Another terminology from graph theory that we need is the definition of *paths*:

Definition 2 A vertex v in a digraph $G = (V, E)$ is *connected* to vertex w via a *path* of length l if there exists a series of edges

$$e_1, e_2, \ldots, e_l \quad e_k \in E \tag{1}$$

where $e_k = (v_{k-1}, v_k)$, $v_0 = v$, $v_l = w$. For undirected graphs, we use the same definition but with $e_k = \{v_{k-1}, v_k\}$.

If such a path exists for some $l > 0$, then i is *connected* to j.

Definition 3 In the context of this work, edges do not have a weight assigned to them, or equivalently, all edges have the weight 1. Hence, there exists a minimal distance between two connected vertices, which is the length of a shortest path between those two. The longest distance between any two vertices in G is called the *diameter* $\text{diam}(G)$ of G.

Remark 1 In *directed* graphs, the relation of connectivity is not symmetric, i.e. vertex v can be connected to vertex w, but at the same time w might not be connected to v.

The notion of points that are connected to each other leads to a global property of the graph G:

Definition 4 Let $G = (V, E)$ be an undirected or directed graph. Then G is called a *strongly connected graph* if for any two vertices $v \in V$ and $w \in V$, v is connected to w and w is connected to v . In addition to that, a directed graph is called *weakly connected* if for any pair of vertices v and w either v is connected to w or w is connected to v.

If a graph G is not connected, it might have *subgraphs* that are connected:

Definition 5 For a graph $G = (V, E)$, a *subgraph* $G' = (V', E')$ consists of a subset of vertices V' together with all edges that are defined on this subset of vertices:

$$E' = \left\{e \in E : v, w \in V'\right\} \tag{2}$$

The combination of Definitions 4 and 5 gives the definition of a *component*:

Definition 6 If a subgraph G' of G is a connected graph, and G' is the largest subgraph with this property, then G' is called a *component* of G.

We can now define graphs $G(A)$ and $\bar{G}(A)$ that are associated with a matrix A in the following sense:

Definition 7 For $A \in \mathbb{R}^{n \times n}$ we can define the *associated graph* $G(A) = (V, E)$ with

$$V = \{i \in \mathbb{N} : 1 \le i \le n\} \text{ and } E = \left\{(i, j) : a_{ij} \neq 0\right\}. \tag{3}$$

In order to represent subsystems that are completely independent from the rest of the linear system, i.e. there are no couplings to or from this subsystem, we need to define the *undirected associated graph* $\bar{G}(A) = (V, \bar{E})$ by using

$$\bar{E} = \left\{\{i, j\} : a_{ij} \neq 0 \vee a_{ji} \neq 0\right\} \tag{4}$$

for the set of edges.

In order to write down the *Depth-First Search* algorithm, we need another definition:

Definition 8 The *adjacency list* $A(v)$ of vertex $v \in V$ in a graph $G(V, E)$ is a list of all vertices $w \in V$ that can be reached from v via a path of length 1.

Now that we have clarified some definitions, let us turn to a basic algorithm that is needed for the detection of components within a graph. Algorithm 1 introduces an algorithm called *Depth-First Search (DFS)*. Starting from some vertex s, the algorithm moves on to one of the neighbors w of v that it has not yet visited. It then

Algorithm 1: Depth-first search (DFS) [18, 19]

```
Procedure MARK = DFS(G,s)
MARK(s) := true;
for all w ∈ A(s) do
    if not MARK(w) then
        DFS(w);
    end if
end for
```

continues to visit an unvisited neighbor of w and so one. Once it reaches a vertex z that has no neighbors that have not been visited, it continues the search at the node from which it has reached z. When the algorithm gets back to s and all neighbors of s have been visited, the algorithm terminates. Algorithm 1 is a recursive version of this method.

Applying DFS to a graph G with starting point s will yield an array MARK where MARK(v) $=$ **true** if and only if v can be reached from s, i.e. if there exists a path from s to v.

Lemma 1 *As, if G is strongly connected, every edge of the graph is touched in the loop, DFS needs* $O(|E|)$ *operations.*

In the following, we will consider *undirected* graphs, if not stated otherwise.

Lemma 2 *If G is an undirected graph, then the subgraph* $G'(V', E')$ *with*

$$V' = \{v \in V : MARK(v) = \textbf{\textit{true}}\} \tag{5}$$

$$E' = \left\{e = \{v, w\} \in E : v, w \in V'\right\} \tag{6}$$

is a component of G.

Proof

1. There is a path from s to every other vertex v in G': By definition, there is a path from s to all $w \in A(s)$. For all $w \in A(s)$ there exist paths to all $w' \in A(w)$ for the same reason. Hence, for all w' we have a path $s \to w \to w'$. By applying this argument recursively, we can find paths from s to all v with MARK(v) $=$ **true**.
2. Since G is undirected, there also exists a path from v to s.
3. Every two other vertices $v, w \in G'$ are connected via a path $v \to \cdots \to s \to \cdots \to w$.

In the case of undirected graphs G we can extend Algorithms 1 and 2 in order to find all components of a graph G. To achieve that, the algorithm needs to re-start the

Algorithm 2: Depth-first search for components (DFS-C) [18]

```
Procedure COMPONENTS = DFS-C(G)
MARK(:) := false
MARK_OLD(:) := MARK(:)
for all v ∈ V do
    MARK := DFS(v)
    for all w ∈ V do
        if MARK(v) ≠ MARK_OLD(v) then
            COMPONENTS(w) := v
        end if
    end for
    MARK_OLD(:) := MARK(:)
end for
```

DFS at every vertex that has not been reached by a previous DFS. When Algorithm 2 (DFS-C) terminates, every vertex has been visited, but the vertices have a label that indicates in which DFS run they have been visited. All vertices that have the same label belong to the same component of G.

If we integrate lines 6–10 from Algorithm 2 into the original DFS algorithm[1] the complexity of DFS-C is $\mathcal{O}(|V| + |E|)$. In arbitrary undirected graphs G, the worst case scenario would be $|E| = |V|(|V| - 1)/2$ which would be the case if every vertex in G was connected to every other vertex. Such graphs are called *dense* [8]. We are mainly interested in graphs associated with sparse matrices arising from discretizations using pointclouds. In this case, $|E|$ depends on the size of the local neighborhoods in the pointcloud, which is significantly smaller than $|V|$. In fact, the neighborhoods N_i in a pointcloud have sizes $\ll 100$ in our applications, i.e. $|E| \approx c|V|$ with $c \ll 100$.

Definition 9 Undirected graphs with

$$|E| \ll |V|(|V| - 1)/2 \tag{7}$$

are called *sparse*.

Lemma 3 *Therefore, for the sparse graphs $\bar{G}(A)$ we are interested in, the complexity of DFS-C is $\mathcal{O}(|V|)$.*

Remark 2 Recall that for $\bar{G}(A)$ the number of vertices $|V|$ is the number of rows in the matrix A so that for these graphs DFS scales linearly in the number of points or matrix rows.

2.2 Related Work

DFS is a serial algorithm and using it in parallel is not straight forward, see for example [7] and more recently [2]. Also note that in FPM every process only holds part of the matrix A and therefore the graph $G(\mathcal{P})$, namely the part that is associated with the part of the pointcloud that resides on that process.

Therefore, methods like the ones proposed in [4, 9] or [14] that assume access to a shared memory cannot be used. The method McColl et al. [9] does have the benefit of being designed specifically for graphs that change over time. Hong et al. [5] consider directed graphs with the *small world* property, i.e. graphs with a small diameter compared to their size. We are mainly interested in undirected graphs that do not have the small world property. The FW-BW method and its extension FW-BW-Trim introduced by Fleischer et al. in [3] and McLendon et al. [10] respectively, implement a divide and conquer strategy. Their drawback is though that after every

[1]Which can be done by using writing the component label directly into the MARK array instead of a binary true/false value.

divide step, the remaining work needs to be redistributed across the participating processors, if a proper load balancing shall be achieved.

2.3 *Origin of Components in FPM*

Now that we have clarified the notations and definitions used in the context of graph theory, let us focus on the question why the linear systems arising in FPM may decompose into multiple smaller linear systems. Section 2.3.1 will explain why the pointcloud may decompose into multiple smaller pointclouds geometrically.

First of all, let us classify the types of components into those that can affect the existence or uniqueness of a solution for the linear system and those that cannot. Components in which both the pressure and the velocity are fixed through appropriate boundary conditions at least one point each are well-posed components, as the corresponding linear systems have a unique solution. This is not the case, if either the velocity or the pressure is not fixed by applying the correct boundary conditions. For example, this would be the case for a component that is confined only by walls, which would mean that the boundary conditions for the pressure all are of Neumann-type. Then, if p is a solution to the pressure in this particular component, so is $p+c$ for a constant c. Note however that this is not the case for the full linear system comprised of all components. From the linear solver perspective it can be crucial to know about the components whose solution is only prescribed up to a constant, see Sect. 2.6.

2.3.1 The Geometric Case

Here we want to point out some situations that produce pointclouds that geometrically induce graphs $\bar{G}(A)$ that decompose into components.

The simplest situation occurs when the simulation itself naturally introduces components because two separate flow domains are being simulated. As an example, consider simulating the flow through a valve that is closing over time. While there is only one flow domain at the beginning when the valve is open, there are two flow domains as soon as the valve has closed. As long as the simulation is set up in a physically correct way, this case leads to well-posed components.

A similar situation occurs when parts of the fluid are separated from the main part because of their velocity, i.e. when droplets of fluid are being formed. The boundary points of such a droplet are detected as free surface boundary points which means we impose Dirichlet boundary conditions for the pressure and free surface boundary conditions for the velocities as described in [17] Section 5.3. Therefore, the linear system is well-posed unless both gravity and inertia tend towards 0, which would be the case for slowly moving droplets in zero gravity. This is a special case however, in which the underlying Incompressible Navier–Stokes equations would admit multiple solutions, that we are not considering with our method.

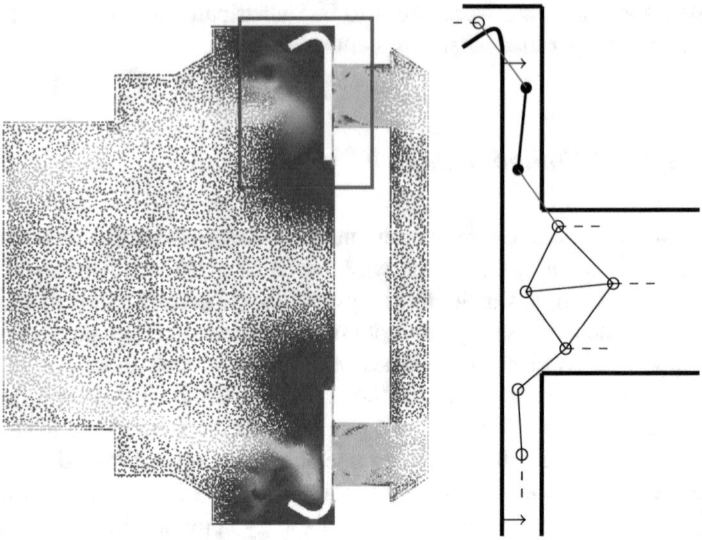

Fig. 1 A valve that is almost closed. The left figure shows the whole valve while the right figure shows a schematic zoomed in on the highlighted area

The most interesting case here is the case where components occur due to a fine detail of the geometry that cannot be resolved properly by the pointcloud. This can happen if the geometry itself is much finer than the resolution of the pointcloud from the beginning, or if the geometry is moving and during this movement fine channels in the geometry are created. An example for this would again be the closing valve. Right before the valve actually closes, there is a very small gap that can be much smaller in width than the average distance between two points of the pointcloud.

Figure 1 shows how a component is formed while the valve is closing. The left part of the figure shows the full valve right before it is fully closed. On the right hand side we see a zoomed in version of the upper right hand side part of the valve shown as a schematic. The lines between two points are indicating that those points are neighbors in the pointcloud. Red lines however indicate that the corresponding points are not connected numerically, e.g. there is no connection between the pressure unknowns in the linear system. In this case here, there is no numerical connection because of the walls intersecting the direct lines between two points. This means that the two points highlighted in bold form a component for themselves. Note that these two points are not detected as free surface boundary points as they have points in multiple directions within their neighborhood. The problem with that is that since both points are considered "interior" points, there are no boundary conditions posed on this small component at all. Therefore, the linear subsystem for this component becomes singular because of the consistency conditions for the least squares problem used to construct the stencils [20]. Analogously there is no numerical coupling between the velocities of these two points and the velocities of the other points in the pointcloud. In Sect. 2.7 we will see that this situation leads to severe problems in some linear solvers when not treated properly.

Situations like this are more of an issue of the discretization method rather than they should be an issue for the linear solver to deal with. We were aiming to improve the performance of a given linear solver (BiCGStab(2)) in this project and BiCGStab(2) did not suffer from the occurring components whereas our AMG method did. Therefore, we saw the need to deal with this situation within our AMG method in order not to break the linear solver in models where it used to work before. Also keep in mind that although detecting components only needs the knowledge of the connectivity graph with respect to the numerical stencils, finding a way to reconnect two components that should only be one component geometrically is a different task. One idea here would be to introduce another point to the pointcloud that serves as a connection between the small and the large component. It is easy to see visually where a point like this should be located, but finding such a point numerically in a 3D problem is not an easy task.

2.3.2 The Algebraic Case

Components in the linear system for solving for the velocity field can also occur even if the linear system for the pressure does not decompose into components. This can be the case when the velocities in the different directions are decoupled from each other because the viscosity η is constant in the entire domain and the boundary conditions also do not impose any couplings between the velocities. These cases are somewhat rare though and will not be the focus of this work.

2.4 *Detecting Components in Parallel*

Let us concentrate on the geometric case of the previous section, i.e. we examine the connectedness of the undirected sparse graph $\bar{G}(A)$ associated the linear system for one of the pressure fields in FPM.

We use variation of an algorithm described by Donev in [1]. It also fits into the framework used by Iverson et al. [6] whereas they use the term *label propagation* for what we will call *local diffusion*.

In order to detect all components in $\bar{G}(A)$, Algorithm 3 is run on all processes involved in parallel. Like the pointcloud, the graph $\bar{G}(A)$ we are looking at is distributed across multiple processes. By V^{loc} we denote those vertices $v \in V$ of the graph that are local to a process. Similarly, E^{loc} denotes all edges $e = \{v, w\}$ for which $v \in V^{\text{loc}}$ and $w \in V^{\text{loc}}$. In contrast to that, E^{remote} denotes edges $e = \{v, w\}$ that satisfy $v \in V^{\text{loc}}$ or $w \in V^{\text{loc}}$. Without loss of generality, in the following we will assume $v \in V^{\text{loc}}$ for $e \in E^{\text{remote}}$. Algorithm 3 yields a label for each vertex indicating to which component it belongs.

Algorithm 3: Parallel detection of components

1: **Procedure** COMPONENTS = **GET-CMP-Par**(G)
2: // Find components locally
3: COMPONENTS := DFS-C$\left(G^{\text{loc}} = \left(V^{\text{loc}}, E^{\text{loc}}\right)\right)$
4: ROOTS:=COMPONENTS
5: // Condense remote edges to "root" of their component
6: **for all** $e = (v, w) \in E^{\text{remote}}$ **do**
7:     $e_R :=$ (COMPONENTS(v), COMPONENTS(w))
8: **end for**
9: // Define reduced graph
10: $E_R^{\text{loc}} := \{e_R\}$
11: $V_R^{\text{loc}} := \left\{v \in V^{\text{loc}} : \text{COMPONENTS}(v) = v\right\}$
12: // Local diffusion
13: **while not** convergence **do**
14:     **for all** $e \in E_R^{\text{loc}}$ **do**
15:         COMPONENTS(v) := min (COMPONENTS(v), COMPONENTS(w))
16:     **end for**
17:     Update COMPONENTS vector
18: **end while**
19: // Update COMPONENTS label in full graph
20: **for all** $v \in V^{\text{loc}} \setminus V_R^{\text{loc}}$ **do**
21:     COMPONENTS(v) := COMPONENTS(ROOTS(v))
22: **end for**

Remark 3 Here we examine the graph $\bar{G}(A)$ rather than the graph $G(A)$. We are interested in independent linear systems within a larger linear system and $\bar{G}(A)$ is a suitable representation for the connectivity of the larger system, see Sect. 2. Although $\bar{G}(A)$ is an undirected graph by the means of Definition 7, in this section we represent each undirected edge $e = \{v, w\}$ through two directed edges $e_1 = (v, w)$ and $e_2 = (w, v)$. This corresponds to the situation that we have when implementing these algorithms in software, where we save adjacency lists for all vertices, in which case an undirected edge is realized in the same fashion.

Remark 4 In this section and specifically in Algorithm 3 we use COMPONENTS as a global array, i.e. we implicitly assume that every processes knows the value of COMPONENTS(i) for every i, even for those vertices that reside on other processes. In our specific implementation this is realized by using a local array for the values of COMPONENTS(i) that correspond to local vertices and another array for the values of non-local vertices that are actually needed. Thus there is no need to have a global, synchronized array COMPONENTS. The update of the latter array needs to be done at the appropriate places in the algorithms. In our description, we omit these updates for simplicity.

The first step in finding all connected components globally is to find all local components on every process first; that is, ignoring all edges that are connections to vertices that reside on other processes. In the following, we will call the latter *remote edges*. Finding the local components (line 3 of Algorithm 3) is done by

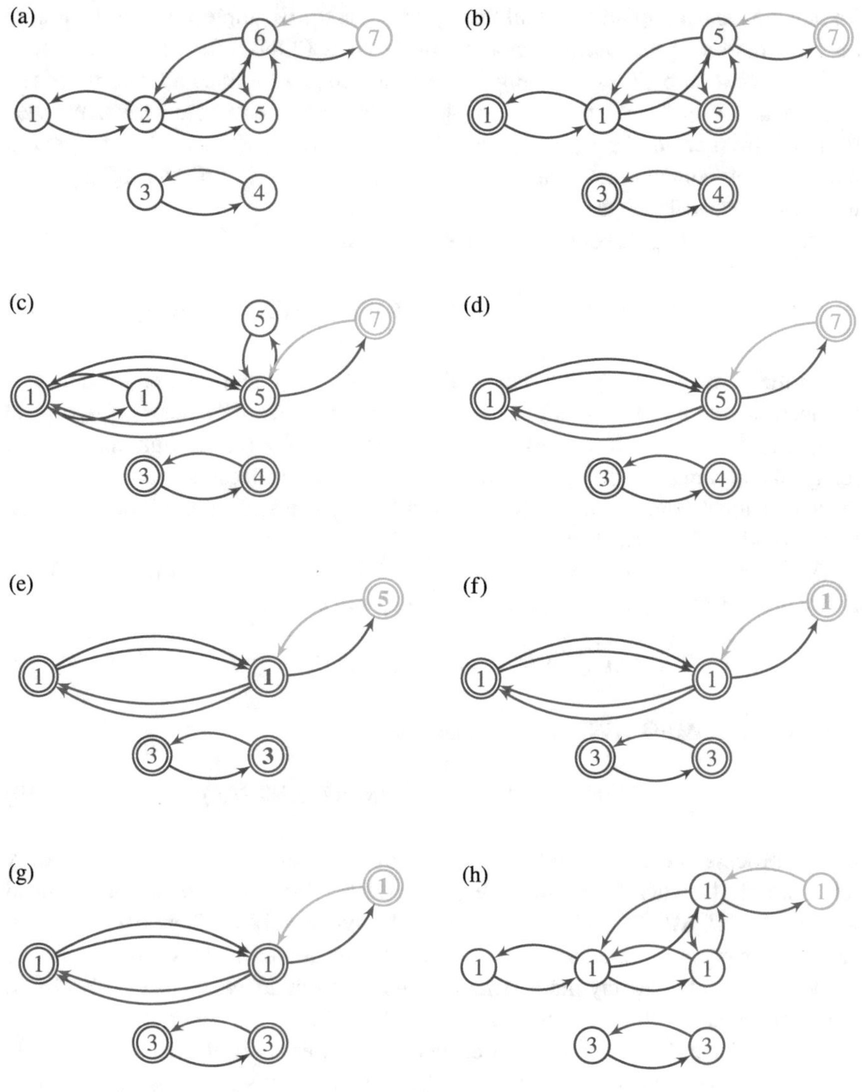

Fig. 2 (**a**) The original graph. (**b**) After detecting local components. Component roots with two circles. (**c**) With remote edges changed to start and end at root vertices. (**d**) The reduced graph. (**e**) The reduced graph after one iteration of local diffusion. Updated values in **bold**. (**f**) The reduced graph after two iterations of local diffusion. Updated values in **bold**. (**g**) The reduced graph after three iterations of local diffusion. No values have been updated, so the algorithm stops. (**h**) After updating the local edges with the new COMPONENTS label

the DFS-C algorithm (see Algorithm 2) introduced in Sect. 2.1. Here, we could also use a different algorithm to detect the local components as long as it has the same asymptotic performance. Figure 2b shows the original graph from Fig. 2a after finding the local components on all three processes, see Example 1. In line 4 we store the information on the local components in the array ROOTS for later reference.

In the next step (lines 6–8; Fig. 2c) the algorithm examines all remote edges. Assume that $e = (v, w)$ is a remote edge where w is the remote vertex. We then want to introduce an edge $e_R = (v', w')$ that connects the two vertices that represent the local components that v and w belong to, i.e. $v' = \text{COMPONENTS}(v)$ and $w' = \text{COMPONENTS}(w)$.

Lines 10–11 (Fig. 2d) define a *reduced graph* with

$$E_R^{\text{loc}} := \{e_R\} \text{ and } V_R^{\text{loc}} := \left\{v \in V^{\text{loc}} : \text{COMPONENTS}(v) = v\right\}, \tag{8}$$

where the vertices are representatives of the local components and edges indicates connections across processes between those components. Obviously, if two local components on different processes are connected in this reduced graph, they are really one component that is spread across those two processes. This means that these two local components should end up having the same global identifier, i.e. they should be detected as one large component.

On the reduced graph, we perform a *local diffusion* algorithm in lines 13–18: Every process checks for every remote edge $e_R = (v, w)$ if

$$\text{COMPONENTS}(w) < \text{COMPONENTS}(v). \tag{9}$$

If (9) holds, $\text{COMPONENTS}(v)$ is updated as

$$\text{COMPONENTS}(v) := \text{COMPONENTS}(w) \tag{10}$$

Every process does this on its local copy of the COMPONENTS array (cf. Remark 4). Therefore, after every process has done the update for its local vertices, the COMPONENTS array needs to be synchronized. Then, (9) is checked again for every remote edge. These steps are repeated until the COMPONENTS array does not change any more. This way, the minimal labels COMPONENTS(u) *diffuse* through the components of the reduced graph, see Fig. 2e–g.

We now have the final labels for all vertices that are part of the reduced graph. The last step is to update the label for all other vertices (lines 20–22 of Algorithm 3). An easy way of doing this is to store from which root vertex a vertex has been visited during DFS-C (line 3 of Algorithm 3). We have already stored that information in the ROOTS array after running DFS-C. Therefore, this final update comes down to setting

$$\text{COMPONENTS}(v) := \text{COMPONENTS}(\text{ROOTS}(v)) \tag{11}$$

for all $v \in V^{\text{loc}} \setminus V_R^{\text{loc}}$. After this, the COMPONENTS array contains the correct label for every $v \in V$.

Remark 5 In the reduced graph, we could remove duplicate edges. Those might exist because two local components on two different processes can be connected via multiple edges from different vertices in both components. For example, in Fig. 2a both the vertex 5 and the vertex 6 that reside on process 2 are connected to vertex 2 on process 1. Because both vertex 6 and vertex 5 belong to the same local component, those two edges are reduced to edges e_R and e'_R connecting vertex 5 and vertex 1 in the reduced graph, see Fig. 2d. One of those edges would be sufficient for the local diffusion part of the algorithm to work correctly, however the second edge does not cause a problem either. The loop in lines 14–16 performs one integer comparison for each remote edge. We would not have to do this comparison for the duplicate remote edges, if we removed them from the reduced graph. However, finding those remote edges would require at least

$$O\left(|E_R^{\mathrm{loc}}| \log |E_R^{\mathrm{loc}}|\right) \overset{|E_R^{\mathrm{loc}}| \leq c |V_R^{\mathrm{loc}}|}{=} O\left(|V_R^{\mathrm{loc}}| \log |V_R^{\mathrm{loc}}|\right) \tag{12}$$

operations for sorting all remote edges.

This would mean that the bound we will show in Sect. 2.5 would not hold any more. Our experiments in Sect. 2.7 also show that the potential gain here is relatively small as we do not perform many iterations of the local diffusion algorithm in practice.

Example 1 The graph depicted in Fig. 2a shall serve as an example for Algorithm 3. The vertex labels correspond to the value of the COMPONENTS label in the current step, whereas the color of each vertex indicates the process on which the vertex resides. Analogously, edges are colored according to the process on which they originate. In some sense, this is a worst case example: The component that has label "1" at the very end (see Fig. 2h) stretches across all three processes and the diffusion of the minimum label "1" needs to pass all three processes before the algorithm stops. Hence in this case we reach the theoretical maximum of P local diffusion iterations that we will derive in Sect. 2.5.

2.5 Complexity of the Algorithm

In this section, we deal with the asymptotic complexity of our proposed algorithm in the case of sparse graphs $\bar{G}(A)$. Performance considerations regarding runtime will be the topic of Sect. 2.7.

Our argument is similar to the one in [6] but because of our more specific knowledge of the graph $\bar{G}(A)$ we can give more specific estimates.

We begin by formulating some estimates regarding the relationships between the number of vertices and the number of edges in various graphs involved here.

Lemma 4

1. $\bigcup V^{loc} = V$, $\sum_p |V^{loc}| = |V|$
2. $|V^{loc}| \leq |V|$, $|E^{loc}| \leq |E|$
3. $|E| \leq c|V|$, $|E^{loc}| \leq c|V^{loc}|$ *with* $c \ll 100$
4. $|V_R^{loc}| \leq |V^{loc}|$

Proof

1. Every point in the pointcloud and therefore every vertex in the graph we are considering here is associated to exactly one process.
2. V^{loc} and E^{loc} are subsets of V and E respectively
3. The number of edges in the graph $\bar{G}(A)$ is limited by the allowed neighborhood size in the pointcloud. In FPM, we usually allow up to 40 neighbors, see for example [16].
4. Vertices in the reduced graph represent local components and there cannot be more local components than local vertices in the original graph.

Lemma 5 *For sparse graphs and a set* **P** *of P processors, Algorithm 3 has an asymptotic complexity of*

$$\mathcal{O}\left(|V| \cdot P\right) \tag{13}$$

in the worst case and

$$\mathcal{O}\left(|V|\right) \tag{14}$$

in the average case.

Proof As we have already seen in Lemma 3, DFS-C has an asymptotic complexity of $\mathcal{O}(|V^{\mathrm{loc}}|)$ on every process. The assignment in line 4 of Algorithm 3 is also of complexity $\mathcal{O}(|V^{\mathrm{loc}}|)$.

Finding E^{remote} which is needed in line 6 can be done in $\mathcal{O}(|E^{\mathrm{loc}})$.[2] Because of Lemma 4, this is also $\mathcal{O}(|V^{\mathrm{loc}}|)$. The loop in lines 6–8 needs E^{remote} iterations which, by the same argument, is also bound by $\mathcal{O}(|V^{\mathrm{loc}}|)$.

Lines 10 and 11 are only notations that are not carried out in software, so we omit those two in our considerations.

For the local diffusion part in lines 13–18 first consider the inner loop in lines 14–16. The number of iterations in this loop is $|E_R^{\mathrm{loc}}|$ and by Lemma 4 we have

$$|E_R^{\mathrm{loc}}| \leq c|V_R| \leq c|V| \quad \text{where } V_R = \bigcup V_R^{\mathrm{loc}}. \tag{15}$$

[2]With an appropriate numbering of the local vertices versus the remote vertices or by labeling those edges beforehand.

It remains to examine how many local diffusion iterations are needed before convergence is reached.

To this end, consider the reduced graph

$$G_R = \left(V_R, \bigcup E_R^{\mathrm{loc}}\right) \tag{16}$$

in which each vertex represents a local component. The components of G_R then yield the global components. In the worst case the minimum label in a component in G_R needs to propagate along the longest shortest path (*diameter*; see Definition 3) in G_R. Afterwards, another iteration is needed to notice that the local diffusion has converged. This makes for a worst case iteration count of

$$I \leq \mathrm{diam}\,(G_R) + 1 \tag{17}$$

Since the vertices in G_R represent local components on each process, there are no edges between two vertices in G_R on the same process. Therefore

$$\mathrm{diam}\,(G_R) \leq P - 1 \tag{18}$$

which yields

$$I \leq P \tag{19}$$

Another assumption for the worst case would be that the vertices are spread unevenly across the processes, i.e.

$$\max_{p \in \mathbf{P}} |V^{\mathrm{loc}}| \approx |V| \tag{20}$$

In this case, all the local complexities in the previous steps of the algorithm become $O(|V|)$ and local diffusion needs P iterations of an $O(|V|)$ loop, meaning the local diffusion part has the highest complexity in this algorithm: $O(|V| \cdot P)$.

For the average case we assume that the vertices are spread evenly across the processors, i.e.

$$\max_{p \in \mathbf{P}} |V^{\mathrm{loc}}| \approx |V|/P \tag{21}$$

Then, the complexities in the local part of the algorithm become $O(|V|/P)$ and the local diffusion part is

$$O(|V|/P \cdot P) = O(|V|). \tag{22}$$

Remark 6 Section 2.7 will show that this theoretical complexity is a pessimistic estimate for many cases. The main bottleneck for the complexity is the number of iterations of the local diffusion part. In the proof above, we have estimated

$$I \leq \operatorname{diam}(G_R) + 1 \leq P \tag{23}$$

Note that the diameter of the reduced graph G_R mainly depends on the partition of the pointcloud onto the processes which in turn depends on the shape of the computational domain. For example, consider a long and thin domain like a tube. In this case, the partition would also follow this shape leading to a graph G_R with a large diameter like $P - 1$ in the worst case. If however the domain is a cube and every process has an equal cubical part of the pointcloud to compute, then

$$\operatorname{diam}(G_R) \approx \sqrt{3}\sqrt[3]{P}, \tag{24}$$

so in this case the complexity of Algorithm 3 would be

$$O\left(|V| \cdot P^{-\frac{2}{3}}\right) \tag{25}$$

2.6 Dealing with Components in the Linear Solver

There are two main types of components that our AMG method distinguishes.

Parallel components that are solved in parallel. In order to determine which components will be solved in parallel, we first find the P largest components. These components are assigned to sets of processors of different sizes according to their size. Specifically, we employ Algorithm 4 for this task.

Serial components that are solved on a single process. Every component that does not belong to the P largest component and every component that has been assigned only 1 processor in Algorithm 4 becomes a serial component. Those serial components that are already located on a single processor will stay on this processor and will be solved by the same processor after it has solved the parallel component it was assigned to. Serial components that reside on more than one processor get redistributed to a single processor in a round-robin fashion. Again, they are solved by their assigned processor after it has solved its parallel component.

Algorithm 4: Algorithm to decide how many processors are assigned to a component

Procedure ASSIGN
$N_{\text{left}} := N$
$P_A := P$
for icmp $\in$ largest_cmp **do**
  ASSIGN(icmp) := $\min\left(\lfloor N_{\text{icmp}}/N_{\text{left}}(P - P_A) + 0.5\rfloor, (P - P_A)\right)$
  $N_{\text{left}} := N_{\text{left}} - N_{\text{icmp}}$
  $P_A := P_A -$ ASSIGN(icmp)
end for

For serial components whose size is below a certain threshold, the default being 100 variables, we do not employ an AMG method but use a direct solver, MKL's PARDISO, right away.

Both PARDISO and our AMG method have special modes to solve a system only up to an additive constant in the case of zero rowsum matrices.

2.7 Numerical Experiments

2.7.1 The Valve Case

Let us turn to the case of a closing valve again, see Sect. 2.3.1. In this real world example, we find that at some point during the simulation, just before the valve closes completely, a small component with 5 points is formed. Note that this is very small compared to the overall size of the pointcloud, which is 380,641. Because of the consistency conditions for the stencil [20], the row sums of the rows in the pressure system for those 5 points are all zero. In this particular model, gravity is disabled which means that the external body forces acting on these points is 0. At the same time, the initial guess is zero, which means that the subsystem has the form

$$\begin{pmatrix} 1 & -\frac{1}{4} & -\frac{1}{4} & -\frac{1}{4} & -\frac{1}{4} \\ -\frac{1}{4} & 1 & -\frac{1}{4} & -\frac{1}{4} & -\frac{1}{4} \\ -\frac{1}{4} & -\frac{1}{4} & 1 & -\frac{1}{4} & -\frac{1}{4} \\ -\frac{1}{4} & -\frac{1}{4} & -\frac{1}{4} & 1 & -\frac{1}{4} \\ -\frac{1}{4} & -\frac{1}{4} & -\frac{1}{4} & -\frac{1}{4} & 1 \end{pmatrix} \begin{pmatrix} 0 \\ 0 \\ 0 \\ 0 \\ 0 \end{pmatrix} = \begin{pmatrix} 0 \\ 0 \\ 0 \\ 0 \\ 0 \end{pmatrix} \tag{26}$$

Hence, the linear system is singular, but the initial guess is already a solution. Consequently, both BiCGStab(2) and Gauss–Seidel solve the full linear system without being affected by the singular subsystem, as both methods do not change the values of the initial guess at the corresponding rows. In the one processor case, their convergence rates are 0.947 and 0.999 respectively, so they need a lot of iterations to solve the system up to the desired accuracy. But this observation is the same without a singular subsystem, see [12].

On the other hand, our AMG method is affected severely by the singular subsystem: Fig. 3 shows that except the for 1 and 128 processor case, the number of AMG iterations is much higher when solving the full linear system, compared to solving the system without the singular subsystem. In order to understand the exceptions of the 1 and 128 processor case, we first need to look at why the other cases need that many iterations.

We find that the coarse level solver is not finding an appropriate solution to the coarse level problem in these cases, because the coarse level problem is singular. This is simply because the singularity on the finest level was transferred down all the way to the coarse level. In the serial case, this does not happen because when constructing the second level, one of the five rows corresponding to the singular

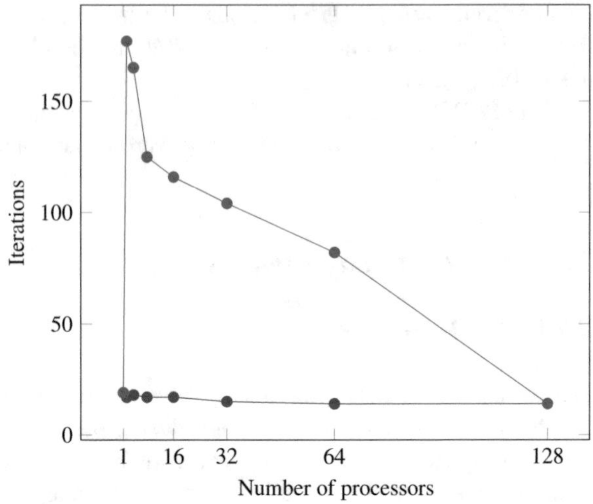

Fig. 3 Number of AMG iterations when working on the full matrix (red) and with the solver being aware of components (blue)

subsystem is picked to be on the second level, while the others automatically become fine level rows, because all the couplings are equal $(-1/4)$ in the stencil at the singular component. Therefore, all the couplings in every row are considered to be strong couplings, see [15], so as soon as one of the five rows is chosen to be on the coarse level, all others have to stay on the fine level. Then on the second level, the former singular component is represented by only one single, independent variable.[3] In this case, our AMG algorithm flags this variable to stay on the second level—as the smoother will solve for this variable directly—before starting to construct the third level. Consequently, the singular component is not represented any more from the third level downwards.

The reason why this is not happening in most of the parallel cases is that in our AMG method we consider couplings that couple rows that reside on different processors to be weak couplings, no matter how large they are by absolute value. I.e. the singular subsystem is not reduced to a single point and then not taken to the next coarse level as in the serial case. Instead, if the singular subsystem resides on more than one processor, it is reduced to a single point on every processor, but it remains on all coarser levels because of its couplings to the other processors. Note that, to this end, the 128 processor case is a special case where the five points comprising the singular component happen to be all on one processor again.

With the component detection turned on, we only solve the main, regular linear system with our AMG method and the small, singular linear system is passed to MKL's direct solver PARDISO, which solves this small system up to a constant.

[3]Note that if following the algorithm described in [15] closely, the diagonal entry in the corresponding row would be 0, leading to a coarse level equation $0 = 0$. Our AMG method implements a check to avoid non-positive diagonals on coarse levels, that causes this entry to be 1 instead.

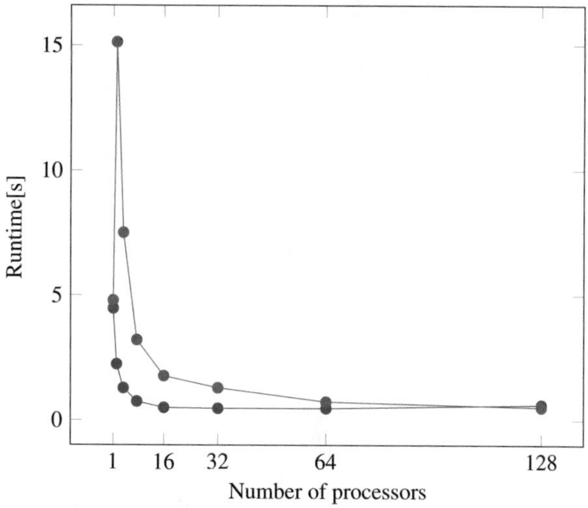

Fig. 4 Runtime when working on the full matrix (red) and with the solver being aware of components (blue)

Note that since in FPM we are only interested in the pressure gradient ∇p anyway, so knowing the solution to the small subsystem up to a constant is sufficient for our purposes. For the performance of our AMG method on the main component, we refer to [11] and [12].

Comparing the run times of both the AMG method with and without the detection of independent components, we see that the benefit from finding the singular component in terms of run time is smaller than the difference in iteration numbers indicates, see Fig. 4. That is because the setup cost when solving the reduced system without the singular component is between 31% and 64% of the overall solver run time and this portion of the run time is the same whether the full system is solved or just the non-singular part. Removing the singular component from the system only speeds up the iteration part of the solving process by reducing the number of iterations needed.

On the other hand though, finding the components and redistributing the system according to Sect. 2.6 requires time that is not needed when solving the full system. Figure 5 shows that the former task is achieved in less than 4% of the overall solver run time. This is better than the theoretical estimate in Lemma 5 suggests. The reason for that is that in Lemma 5 we assumed that the number of iterations of the local diffusion algorithm can be up to $P + 1$, where P is the number of processes involved.

However, Fig. 6 shows that even though the number of iterations is increasing with the number of processes used, for 1024 processes, still only 9 iterations are needed. This observation means that the local diffusion part of Algorithm 3 is actually a lot less expensive than predicted and that Remark 6 is important here. It is a *best case* addition to Lemma 5 and although the actual number of iterations depends on the geometry of the given problem, many results are closer to the best case than to the average case.

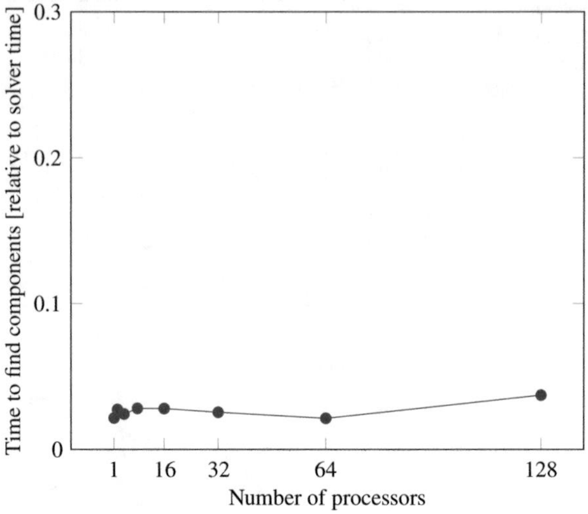

Fig. 5 Time needed to determine the component structure of the graph relative to the overall time needed to solve the system

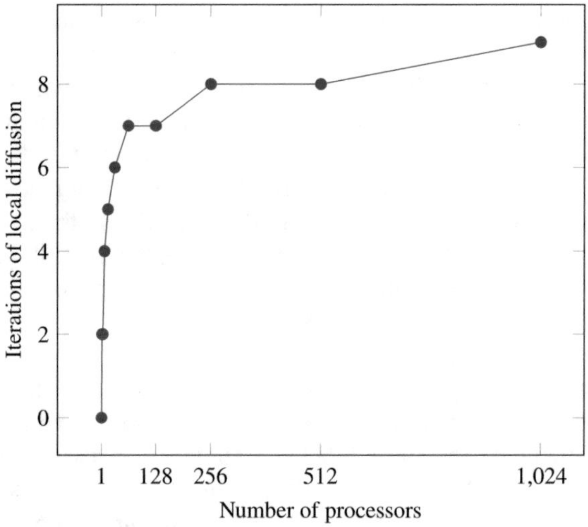

Fig. 6 Number of iterations of local diffusion for different numbers of processors

After the components have been found, we need to redistribute the system according to Sect. 2.6. This task is much more expensive in terms of run time, as Fig. 7 shows. In other words, while finding the components only takes less than 4% of the overall run time, redistributing the matrix takes between 20% and 30%.

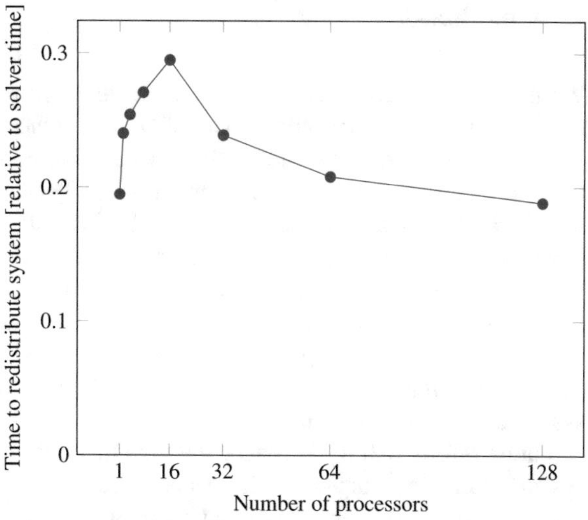

Fig. 7 Time needed to redistribute the matrix across the processors relative to the overall time needed to solve the system

3 Avoiding Redistribution

The results in Sect. 2 indicate that finding the components comes at a negligible cost compared to redistributing the linear system. Thus, future work will be dedicated to finding means to avoid or at least reduce the amount of redistribution that needs to be done.

For the case of one large component and one or several very small ones, like in the example of the closing valve, it would be beneficial to join the small components with the large one but make sure that the variables corresponding to the small, potentially singular, components stay on the finest level in the AMG hierarchy. This way, they would only be affected by the smoother and the Krylov accelerator which would mean they would be solved by a method comparable to the status quo (BiCGStab(2)). In the case shown in the previous section, this method would have worked and the redistribution could have been avoided altogether as only one component would have been left, which our AMG method then would have solved with the given partition across the processes. A variant of this method would be to also add some large value to diagonal entries corresponding to the small components. This would make sure that the small components do not affect the convergence of the AMG method dramatically. The solution to these components could then be computed separately after the main AMG method. Theoretically this would come done to ignoring the small components and solving the large one with AMG without redistributing it before and then solving the small components. A third option is to treat the small components as a Schwarz block without any couplings to the remainder block within the large component. Again, this would remove the small components from the AMG method and solve them separately without a need to redistribute the full linear system. Similarly, restricting the redistribution to small components and solving the parallel component with the given partition is an option.

4 Conclusions

We have introduced an algorithm that can find components in a sparse graph in parallel with a theoretical complexity of $O(|V|)$ in the average case.

Our experiments have shown though that this is a fairly rough estimate and that the measured performance is much better, because the local diffusion needs less iterations than predicted theoretically. Overall, the time needed to detect components is below 4% of the overall solver run time in the example shown here.

In the simulation of a closing valve, where a singular component occurs right before the valve closes completely, our AMG method has shown a stable convergence across multiple different numbers of processors when detecting and treating the singular component separately, whereas the AMG method on the full system showed a very unstable behavior.

Future efforts will try to reduce the time that is needed to redistribute the linear system after the components have been found. Another option is to avoid the redistribution completely, as pointed out in Sect. 3.

References

1. A. Donev, Connected components of a graph. http://computation.pa.msu.edu/NO/ConnCompPresentation.html. Accessed 16 Feb 2018
2. J.A. Edwards, U. Vishkin, Better speedups using simpler parallel programming for graph connectivity and biconnectivity, in *Proceedings of the 2012 International Workshop on Programming Models and Applications for Multicores and Manycores* (ACM, New York, 2012), pp. 103–114
3. L.K. Fleischer, B. Hendrickson, A. Pınar, On identifying strongly connected components in parallel, in *International Parallel and Distributed Processing Symposium* (Springer, Berlin, 2000), pp. 505–511
4. D.S. Hirschberg, A.K. Chandra, D.V. Sarwate, Computing connected components on parallel computers. Commun. ACM **22**(8), 461–464 (1979)
5. S. Hong, N.C. Rodia, K.Olukotun, On fast parallel detection of strongly connected components (SCC) in small-world graphs, in *Proceedings of the International Conference on High Performance Computing, Networking, Storage and Analysis* (ACM, New York, 2013), p. 92
6. J. Iverson, C. Kamath, G. Karypis, Evaluation of connected-component labeling algorithms for distributed-memory systems. Parallel Comput. **44**, 53–68 (2015)
7. V. Kumar, V. Nageshwara Rao, Parallel depth first search. Part II. Analysis. Int. J. Parallel Program. **16**(6), 501–519 (1987)
8. L. Lovasz, B. Szegedy, Limits of dense graph sequences. J. Comb. Theory Ser. B **96**(6), 933–957 (2006)
9. R. McColl, O. Green, D.A. Bader, A new parallel algorithm for connected components in dynamic graphs, in *20th Annual International Conference on High Performance Computing* (IEEE, Piscataway, 2013), pp. 246–255
10. W. McLendon III, B. Hendrickson, S.J. Plimpton, L. Rauchwerger, Finding strongly connected components in distributed graphs. J. Parallel Distrib. Comput. **65**(8), 901–910 (2005)
11. B. Metsch, F. Nick, J. Kuhnert, Algebraic multigrid for the finite pointset method. Comput. Vis. Sci. (reviewed)

12. F. Nick, B. Metsch, H.-J. Plum, Linear solvers for the finite pointset method, in *Recent Advances in Computational Engineering*, ed. by M. Schäfer, M. Behr, M. Mehl, B. Wohlmuth (Springer, Cham, 2018), pp. 89–110
13. G.L.G. Sleijpen, D.R. Fokkema, Bicgstab(l) for linear equations involving unsymmetric matrices with complex spectrum. Electron. Trans. Numer. Anal. **1**, 11–32 (1993)
14. G.M. Slota, S. Rajamanickam, K. Madduri, Bfs and coloring-based parallel algorithms for strongly connected components and related problems, in *IPDPS '14 Proceedings of the 2014 IEEE 28th International Parallel and Distributed Processing Symposium* (IEEE, Piscataway, 2014), pp. 550–559
15. K. Stüben, An introduction to algebraic multigrid, in *Multigrid* (Academic, London, 2000)
16. P. Suchde, Conservation and accuracy in meshfree generalized finite difference methods, Ph.D. thesis, University of Kaiserslautern, 2017
17. P. Suchde, J. Kuhnert, S. Tiwari, On meshfree GFDM solvers for the incompressible Navier–Stokes equations. Comput. Fluids **165**, 1–12 (2018)
18. R. Tarjan, Depth-first search and linear graph algorithms. SIAM J. Comput. **1**(2), 146–160 (1972)
19. R. Tarjan, Finding dominators in directed graphs. SIAM J. Comput. **3**(1), 62–89 (1974)
20. S. Tiwari, J. Kuhnert, Finite pointset method based on the projection method for simulations of the incompressible Navier–Stokes equations, in *3rd International Workshop on Meshfree Methods for Partial Differential Equations* (Springer, Berlin, 2002)

Numerical Study of the RBF-FD Level Set Based Method for Partial Differential Equations on Evolving-in-Time Surfaces

Andriy Sokolov, Oleg Davydov, and Stefan Turek

Abstract In this article we present a Radial Basis Function (RBF)-Finite Difference (FD) level set based method for the numerical solution of partial differential equations (PDEs) of the reaction-diffusion-convection type on an evolving-in-time hypersurface $\Gamma(t)$. In a series of numerical experiments we study the accuracy and robustness of the proposed scheme and demonstrate that the method is applicable to practical models.

1 Introduction

Numerical simulation of partial differential equations posed on an evolving-in-time hypersurface $\Gamma(t)$ is a rapidly growing branch of numerical mathematics, which finds its applications in many industrial tasks. During the last decade many profound finite-element-based methods for surface-defined PDEs were developed: parametric methods [7–9], bulk-layer methods of the phase-field [21] and level-set [8, 23] types, the trace FEM [19] and the space-time FEM [20], etc. All these methods are of the finite element nature, meaning that one has to construct a mesh before any numerical simulation begins. Very often, some largely CPU- and time-consuming work has to be done with or related to the mesh during the simulation process.

On the other hand, kernel methods based on radial basis functions are becoming increasingly popular for the numerical simulation of partial differential equations due to their flexibility of working with scattered data nodes, high accuracy, and significantly simpler implementation. These methods demonstrated promising

A. Sokolov (✉) · S. Turek
Institute for Applied Mathematics, TU Dortmund, Dortmund, Germany
e-mail: andriy.sokolov@math.tu-dortmund.de

O. Davydov
Department of Mathematics, University of Giessen, Giessen, Germany
e-mail: oleg.davydov@math.uni-giessen.de

M. Griebel, M. A. Schweitzer (eds.), *Meshfree Methods for Partial Differential Equations IX*, Lecture Notes in Computational Science and Engineering 129,
https://doi.org/10.1007/978-3-030-15119-5_7

results for various problems of PDEs in two- and three-dimensional domains, see, e.g. [1, 2, 10, 18].

In the recently appeared works of G. Wright et al. [13, 22] the RBF-FD methodology was applied to the simulation of surface PDEs of reaction-diffusion type on stationary manifolds. In the current paper, with the help of the level set technique, we extend the RBF-FD method to reaction-diffusion-convection partial differential equations on evolving-in-time surfaces.

2 PDE on Evolving Hypersurface

2.1 Problem Formulation

We consider the following reaction-diffusion-convection equation

$$\frac{\partial^* u}{\partial t} + \boldsymbol{w} \cdot \nabla_{\Gamma(t)} u = D \Delta_{\Gamma(t)} u + g(u) \;\text{ on }\; \Gamma(t) \times T. \tag{1}$$

Here, the constant D is the viscousity coefficient. $\Gamma(t)$ is a compact, smooth, connected and closed hypersurface in $\mathbb{R}^d$, $d = 2, 3$; $\frac{\partial^* u}{\partial t}$ is a time-derivative, which takes into account the evolution of $\Gamma(t)$ and will be explained below, $\Delta_{\Gamma(t)} u$ is the Laplace-Beltrami term, $\boldsymbol{w}$ is some vector field which transports u along $\Gamma(t)$ and $g(\cdot)$ is a kinetic term. The corresponding initial and boundary (if any) conditions for u have to be provided. We adopt the notation by writing vector fields in bold letters, i.e., $\boldsymbol{c} = (c_1, \ldots, c_n)^T$. We assume that the solution u of (1) can be (naturally) extended from $\Gamma(t)$ to an ϵ-band $\Omega_\epsilon(t)$, see Fig. 1. The domain of interest or also the calculational domain is $\Omega = \Omega_{in} \cup \Omega_{out} \cup \Gamma$. For the sake of simplicity, we also assume that $\Gamma(t) \subset \Omega_\epsilon(t) \subset \Omega$ during the whole simulation time $t \in [0, T]$. The surface derivative

$$\frac{\partial^* u}{\partial t} = \partial_t^\bullet u + u \nabla_{\Gamma(t)} \cdot \boldsymbol{v}$$

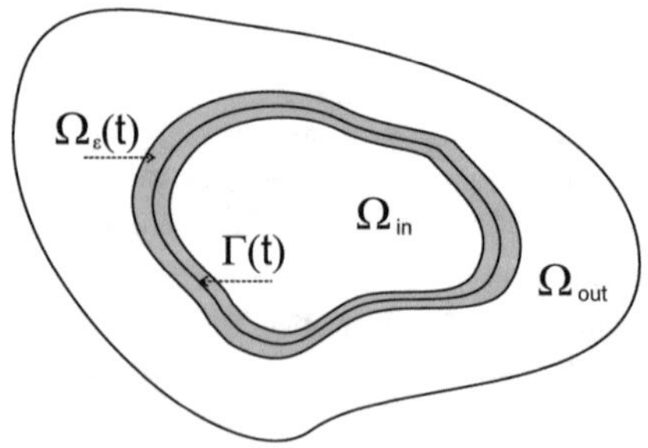

Fig. 1 Geometric illustration

can be obtained by the Leibniz formula

$$\frac{d}{dt}\int_{\Gamma(t)} u = \int_{\Gamma(t)} \partial_t^\bullet u + u\nabla_{\Gamma(t)} \cdot \boldsymbol{v}.$$

By $\partial_t^\bullet u = \partial_t u + \boldsymbol{v} \cdot \nabla u$ one denotes the covariant or advective surface material derivative. The surface velocity $\boldsymbol{v} = V\boldsymbol{n} + \boldsymbol{v}_S$ can be decomposed into velocity components in the normal direction $V\boldsymbol{n}$, with $\boldsymbol{n}$ to be a surface outward normal vector, and in the tangential direction $\boldsymbol{v}_S$. Using the relation

$$\begin{aligned}\nabla_\Gamma \cdot \boldsymbol{v} = \nabla_\Gamma V \cdot \boldsymbol{n} + V\nabla_\Gamma \cdot \boldsymbol{n} + \nabla_\Gamma \cdot \boldsymbol{v}_S = V\nabla_\Gamma \cdot \boldsymbol{n} + \nabla_\Gamma \cdot \boldsymbol{v}_S =\\ = -VH + \nabla_\Gamma \cdot \boldsymbol{v}_S,\end{aligned}$$

and therefore

$$\boldsymbol{v} \cdot \nabla u = V\boldsymbol{n} \cdot \nabla u + \boldsymbol{v}_S \cdot \nabla u = V\frac{\partial u}{\partial \boldsymbol{n}} + \boldsymbol{v}_S \cdot \nabla u,$$

where H is a mean curvature, we can rewrite (1) as

$$\partial_t u + \boldsymbol{v}_S \cdot \nabla u - VHu + V\frac{\partial u}{\partial \boldsymbol{n}} + u\nabla_{\Gamma(t)} \cdot \boldsymbol{v}_S + \boldsymbol{w} \cdot \nabla_{\Gamma(t)} u = D\Delta_{\Gamma(t)} u + g(u),$$

or, in terms of the surface material derivative, as

$$\partial_t^\bullet u + u\nabla_\Gamma \cdot \boldsymbol{v} + \boldsymbol{w} \cdot \nabla_{\Gamma(t)} u = D\Delta_{\Gamma(t)} u + g(u). \tag{2}$$

2.2 Level Set Method

For the implicit prescription of a compact, smoothly connected and oriented hypersurface $\Gamma(t) \subset \Omega$ we introduce a smooth level set function

$$\phi(t, \boldsymbol{x}) = \begin{cases} < 0, & \text{if } \boldsymbol{x} \text{ is inside } \Gamma(t), \\ = 0, & \text{if } \boldsymbol{x} \in \Gamma(t), \\ > 0, & \text{if } \boldsymbol{x} \text{ is outside } \Gamma(t), \end{cases} \tag{3}$$

such that $|\nabla\phi| \neq 0$. Then, an outward normal to $\Gamma(t)$ at the point $\boldsymbol{x}$ is

$$\boldsymbol{n}(\boldsymbol{x}) = (n^1, n^2, \ldots, n^d)^T = \nabla\phi(\boldsymbol{x})/|\nabla\phi(\boldsymbol{x})| \tag{4}$$

and the matrix

$$\mathcal{P}_\Gamma = I - \boldsymbol{n}\boldsymbol{n}^T = \left(\delta_{ij} - n^i n^j\right)_{i,j=1}^d \tag{5}$$

is the projection onto the tangent space $\mathcal{T}_{\boldsymbol{x}}\Gamma(t)$. For a scalar function η and a tangential vector field $\boldsymbol{\eta} = (\eta^1, \eta^2, \ldots, \eta^d)^T$ on Γ extended into Ω we can define

$$\nabla_\Gamma \eta := (\mathcal{P}_\Gamma \nabla)\, \eta = \left\{ \frac{\partial \eta}{\partial x_i} - \sum_{j=1}^{d} n^i n^j \frac{\partial \eta}{\partial x^j} \right\}_{i=1}^{d}, \tag{6}$$

$$\nabla_\Gamma \cdot \boldsymbol{\eta} = \sum_{i=1}^{d} \left(\frac{\partial \eta^i}{\partial x^i} - \sum_{j=1}^{d} n^i n^j \frac{\partial \eta^i}{\partial x^j} \right), \tag{7}$$

the surface gradient ∇_Γ and the surface divergence $\nabla_\Gamma \cdot$ operators, respectively. Using this notation, the Laplace-Beltrami operator can be written as

$$\Delta_\Gamma \eta = \nabla_\Gamma \cdot \nabla_\Gamma \eta = \mathcal{P}_\Gamma \nabla \cdot \mathcal{P}_\Gamma \nabla\, \eta. \tag{8}$$

2.3 *Discretization in Time*

For the discretisation in time of the surface PDE

$$\partial_t u + \boldsymbol{v} \cdot \nabla u + u \nabla_{\Gamma(t)} \cdot \boldsymbol{v} + \boldsymbol{w} \cdot \nabla_{\Gamma(t)} u = D \Delta_{\Gamma(t)} u + g(u), \tag{9}$$

we use the θ-scheme method. Given u^n and the time step $\Delta t = t_{n+1} - t_n$, solve for $u = u^{n+1}$ (for the sake of simplicity we omit the index $\{n+1\}$ there it is possible, e.g. $t = t^{n+1}$)

$$\begin{aligned} \frac{u - u^n}{\Delta t} &+ \theta \left(\boldsymbol{v} \cdot \nabla u + u \nabla_{\Gamma(t)} \cdot \boldsymbol{v} + \boldsymbol{w} \cdot \nabla_{\Gamma(t)} u - D \Delta_{\Gamma(t)} u + g(u) \right) \\ &= -(1-\theta) \left(\boldsymbol{v}^n \cdot \nabla u^n + u^n \nabla_{\Gamma(t^n)} \cdot \boldsymbol{v}^n \right. \\ &\left. + \ \boldsymbol{w}^n \cdot \nabla_{\Gamma(t^n)} u^n - D \Delta_{\Gamma(t^n)} u^n + g(u^n) \right). \end{aligned} \tag{10}$$

If we denote corresponding discrete operators, whose RBF-FD construction will be described in Sect. 3, by

$$L(t, \Gamma(t))\boldsymbol{u} \approx -\Delta_{\Gamma(t)} u|_X, \tag{11}$$

$$\widetilde{K}(t, \boldsymbol{v})\, \boldsymbol{u} \approx -\boldsymbol{v} \cdot \nabla u|_X, \tag{12}$$

$$\widetilde{\widetilde{K}}(t, \boldsymbol{w}, \Gamma(t))\, \boldsymbol{u} \approx -\boldsymbol{w} \cdot \nabla_{\Gamma(t)} u|_X, \tag{13}$$

$$G(t, \Gamma(t))\, \boldsymbol{u} \approx u \nabla_{\Gamma(t)} \cdot \boldsymbol{v}|_X, \tag{14}$$

where $\boldsymbol{u} = (u_1, u_2, \ldots, u_N)^T \approx u|_X = \left(u(\boldsymbol{x}_1), u(\boldsymbol{x}_2), \ldots, u(\boldsymbol{x}_N)\right)^T$, with $X = \{\boldsymbol{x}_j\}_{j=1}^N \subset \Omega$, then the semi-discrete equation (10) can be rewritten in the following matrix form:

$$[\, I + \theta \Delta t \left\{ -\widetilde{K}(t, \boldsymbol{v}) - \widetilde{\widetilde{K}}(t, \boldsymbol{w}, \Gamma) + L(t, \Gamma) + G(t, \Gamma) \right\}]\, \boldsymbol{u}$$

$$= [\, I - (1-\theta)\Delta t \left\{ -\widetilde{K}(t^n, \boldsymbol{v}^n) - \widetilde{\widetilde{K}}(t^n, \boldsymbol{w}^n, \Gamma^n) \right.$$

$$\left. + L(t^n, \Gamma^n) + G(t^n, \Gamma^n) \right\}]\, \boldsymbol{u}^n$$

$$+ \theta \Delta t\, g(\boldsymbol{u}) + (1-\theta)\Delta t\, g(\boldsymbol{u}^n), \tag{15}$$

For our numerical simulations we take either the Implicit-Euler scheme, which corresponds to $\theta = 1$, or the Crank-Nicolson scheme, which is obtained from (15) by setting $\theta = \frac{1}{2}$.

3 RBF-FD for PDEs on Evolving-in-Time Surfaces

3.1 Kernel Interpolation and Operator Approximation

Given a set of scattered nodes $X = \{\boldsymbol{x}_j\}_{j=1}^N \subset \Omega$ we are looking for a continuous function $u : \Omega \to \mathbb{R}$ as a kernel interpolant, those general form is

$$I_\phi u(\boldsymbol{x}) = \sum_{j=1}^{N} c_j \Phi(\boldsymbol{x}, \boldsymbol{x}_j), \quad \boldsymbol{x} \in \Omega, \tag{16}$$

such that its restriction $u|_{\Gamma(t)}$ is a solution of equation (2). Here, Φ is a positive definite kernel called a *radial basis function* (RBF) with the property $\Phi(\boldsymbol{x}, \boldsymbol{y}) = \varphi(\|\boldsymbol{x} - \boldsymbol{y}\|)$. Denoting $\boldsymbol{r}_j(\boldsymbol{x}_i) = \|\boldsymbol{x}_i - \boldsymbol{x}_j\|$, the interpolation coefficients $\{c_j\}_{j=1}^N$ are determined by enforcing $I_\varphi u|_X = u|_X$ as the following linear system:

$$A_X c_X = u_X, \tag{17}$$

where

$$A_X = \begin{bmatrix} \varphi(\boldsymbol{r}_1(\boldsymbol{x}_1)) & \varphi(\boldsymbol{r}_2(\boldsymbol{x}_1)) & \ldots & \varphi(\boldsymbol{r}_N(\boldsymbol{x}_1)) \\ \varphi(\boldsymbol{r}_1(\boldsymbol{x}_2)) & \varphi(\boldsymbol{r}_2(\boldsymbol{x}_2)) & \ldots & \varphi(\boldsymbol{r}_N(\boldsymbol{x}_2)) \\ \vdots & \vdots & \ddots & \vdots \\ \varphi(\boldsymbol{r}_1(\boldsymbol{x}_N)) & \varphi(\boldsymbol{r}_2(\boldsymbol{x}_N)) & \ldots & \varphi(\boldsymbol{r}_N(\boldsymbol{x}_N)) \end{bmatrix}, \quad c_X = \begin{bmatrix} c_1 \\ c_2 \\ \vdots \\ c_N \end{bmatrix}, \quad u_X = \begin{bmatrix} u(\boldsymbol{x}_1) \\ u(\boldsymbol{x}_2) \\ \vdots \\ u(\boldsymbol{x}_N) \end{bmatrix}.$$

For a positive definite φ, this system is positive definite and hence solvable.

In the following we use the radial basis function finite difference (RBF-FD) method for approximation of all linear differential operators, which arise through our derivations. Let $\mathcal{L}$ be one of these linear operators. Then the approximation of $\mathcal{L}u$ at the point $\boldsymbol{\zeta}$ is sought as a weighted sum of function values $u(\boldsymbol{\xi}_j)$ at the points $\Xi = \Xi_\zeta = \{\boldsymbol{\xi}_1, \boldsymbol{\xi}_2, \dots, \boldsymbol{\xi}_K\}$ neighboring to $\boldsymbol{\zeta}$:

$$\mathcal{L}u(\boldsymbol{\zeta}) \approx \sum_{j=1}^{K} \omega_j u(\boldsymbol{\xi}_j), \quad \boldsymbol{\xi}_j \in \Xi, \tag{18}$$

where the approximation weights $\boldsymbol{\omega} = (\omega_1, \omega_2, \dots, \omega_K)^T$ can be computed by solving the linear system

$$A_\Xi \boldsymbol{\omega} = [\mathcal{L}\varphi(\boldsymbol{r}_j(\boldsymbol{\zeta}))]_{j=1}^K \quad \text{with} \quad A_\Xi := [\varphi(\boldsymbol{r}_j(\boldsymbol{\xi}_i))]_{i,j=1}^K. \tag{19}$$

In general, a good choice of stencil points $\boldsymbol{\xi}_i$ for the accurate approximation of $\mathcal{L}u(\boldsymbol{\zeta})$ is a nontrivial task which requires additional analysis [3, 5, 18]. In this article, the set Ξ_ζ consists of the $K = 9$ points nearest to $\boldsymbol{\zeta}$ in the Euclidean distance, including $\boldsymbol{\zeta}$ itself. Either Gaussian $\varphi(r) = \exp(-\epsilon^2 r^2)$ with $\epsilon > 0$ close to zero, or the polyharmonic radial basis function $\varphi(r) = r^\gamma$ with $\gamma = 5$ are used in all presented numerical simulations. In the case of Gaussian we use a QR preconditioning technique that allows stable computation of the weights for any value of the shape parameter ϵ [4, 12, 17]. Polyharmonic RBF is only conditionally positive definite and therefore the interpolant (16) is extended in this case by a polynomial term of degree $\lfloor \gamma/2 \rfloor$, see [10, 11] for details.

In the case of a vector-valued operator $\mathcal{L}$ the weights ω_j are vectors, and $\boldsymbol{\omega}$ is a matrix. In particular, (18) is replaced by

$$\nabla u(\boldsymbol{\zeta}) \approx \boldsymbol{\omega}_\nabla(\boldsymbol{\zeta}, \Xi)^T u_\Xi \tag{20}$$

for the gradient operator ∇, where each column of the matrix $\boldsymbol{\omega}_\nabla(\boldsymbol{\zeta}, \Xi) \in \mathbb{R}^{K\times d}$ is obtained by solving (19) for the corresponding partial derivative operator. Clearly, a gradient-type operator $\mathcal{L}_{\text{grad}}^A u = A\nabla u$ with components $\sum_{j=1}^d a_{ij} \frac{\partial u}{\partial x_j}$, $i = 1, \dots, d$, where $A : \Omega \to \mathbb{R}^{d\times d}$, can be discretized as

$$\mathcal{L}_{\text{grad}}^A u(\boldsymbol{\zeta}) \approx \left[\sum_{i=1}^K \omega_{ij} u(\boldsymbol{\xi}_i)\right]_{j=1}^d = A(\boldsymbol{\zeta})\boldsymbol{\omega}_\nabla^T(\boldsymbol{\zeta}, \Xi) u_\Xi, \tag{21}$$

where $\boldsymbol{\omega} := \boldsymbol{\omega}_\nabla(\boldsymbol{\zeta}, \Xi) A^T(\boldsymbol{\zeta})$. A simple calculation shows that the same weight matrix $\boldsymbol{\omega} = \boldsymbol{\omega}_\nabla(\boldsymbol{\zeta}, \Xi) A^T(\boldsymbol{\zeta})$ gives a discretization

$$\mathcal{L}_{\text{div}}^A \boldsymbol{u}(\boldsymbol{\zeta}) \approx \sum_{i=1}^K \sum_{j=1}^d \omega_{ij} u_j(\boldsymbol{\xi}_i) = \operatorname{trace}\left(A(\boldsymbol{\zeta})\boldsymbol{\omega}_\nabla^T(\boldsymbol{\zeta}, \Xi)\boldsymbol{u}_\Xi\right) \tag{22}$$

of the divergence-type operator $\mathcal{L}^A_{\text{div}}\boldsymbol{u} = A\nabla \cdot \boldsymbol{u} := \sum_{i,j=1}^d a_{ij}\frac{\partial u_i}{\partial x_j}$, where $\boldsymbol{u} = (u_1, \dots, u_d)^T$ is a vector-function, and $\boldsymbol{u}_\Xi = [u_j(\boldsymbol{\xi}_i)]_{i,j=1}^{K,d}$.

Formulas (21) and (22) can be combined to obtain an approximation of the anisotropic diffusion operator

$$\Delta^{A,B}u := A\nabla \cdot B\nabla u = \mathcal{L}^A_{\text{div}}\mathcal{L}^B_{\text{grad}}u, \quad A, B : \Omega \to \mathbb{R}^{d\times d}.$$

To this end, an auxiliary set of points $\Gamma = \{\boldsymbol{\gamma}_1, \dots, \boldsymbol{\gamma}_L\}$ is chosen in the neighborhood of $\boldsymbol{\zeta}$, an approximation of the vector

$$\boldsymbol{u}(\boldsymbol{\gamma}_s) := \mathcal{L}^B_{\text{grad}}u(\boldsymbol{\gamma}_s) \approx \Big[\sum_{i=1}^K \omega_{ij}(\boldsymbol{\gamma}_s)u(\boldsymbol{\xi}_i)\Big]_{j=1}^d, \quad \boldsymbol{\omega}(\boldsymbol{\gamma}_s) := \boldsymbol{\omega}_\nabla(\boldsymbol{\gamma}_s, \Xi)B^T(\boldsymbol{\gamma}_s),$$

is obtained by (21) for each $s = 1, \dots, L$, and inserted into (22), where Γ is used instead of Ξ. Setting $\tilde{\boldsymbol{\omega}} := \boldsymbol{\omega}_\nabla(\boldsymbol{\zeta}, \Gamma)A^T(\boldsymbol{\zeta})$, we arrive at

$$\Delta^{A,B}u(\boldsymbol{\zeta}) \approx \sum_{i=1}^K \omega_i u(\boldsymbol{\xi}_i), \quad \omega_i = \sum_{s=1}^L\sum_{j=1}^d \tilde{\omega}_{sj}\omega_{ij}(\boldsymbol{\gamma}_s), \tag{23}$$

that is

$$\omega_i = \text{trace}\left(\tilde{\boldsymbol{\omega}}\,[\omega_{ij}(\boldsymbol{\gamma}_s)]_{j,s=1}^{d,L}\right), \quad i = 1, \dots, K.$$

In the case when $A = B$ and $\boldsymbol{\zeta} \in \Gamma = \Xi$ the formulas for ω_i in (23) can be simplified since $\tilde{\boldsymbol{\omega}}$ coincides with one of the matrices $\boldsymbol{\omega}(\boldsymbol{\gamma}_s)$, see [13, 22]. We however prefer to choose Γ closer to $\boldsymbol{\zeta}$, in order to obtain more reliable numerical differentiation formulas for $\mathcal{L}^B_{\text{grad}}u(\boldsymbol{\gamma}_s)$. In this paper we use

$$\boldsymbol{\gamma}_j = (\boldsymbol{\zeta} + \boldsymbol{\xi}_j)/2, \quad j = 1, \dots, K, \tag{24}$$

see Fig. 2, where $\boldsymbol{\xi}_1 = \boldsymbol{\zeta}$.

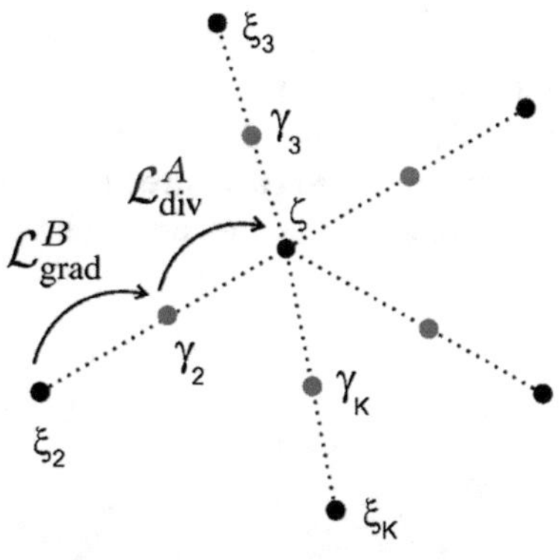

Fig. 2 Discretization of the anisotropic diffusion operator

3.2 RBF-FD Discretization in Space

We now describe the discrete operators in (11)–(14). After choosing a set of nodes $X = \{\boldsymbol{x}_j\}_{j=1}^N \subset \Omega$, we select for each $\boldsymbol{\zeta} \in X$ a set of neighbors $\Xi_{\boldsymbol{\zeta}} \subset X$.

Thanks to (8), the value of the Laplace-Betrami operator $-\Delta_{\Gamma(t)} u(\boldsymbol{\zeta})$ can be approximated according to (23) with $-A = B = \mathcal{P}_{\Gamma(t)}$, and the weights $\boldsymbol{\omega}_i$ of this formula become the nonzero entries of the $\boldsymbol{\zeta}$-row of the matrix $L(t, \Gamma(t))$ in (11).

For the generalized RBF-FD approximation of convection operators $\boldsymbol{v} \cdot \nabla u$ and $\boldsymbol{w} \cdot \nabla_{\Gamma(t)} u$ we make an assumption that both vector fields $\boldsymbol{v}$ and $\boldsymbol{w}$ can be extended outside of $\Gamma(t)$ to the whole domain Ω. In the case of the level set framework this extension of the surface velocity $\boldsymbol{v}$ is straightforward as a velocity field of the corresponding level set. Then by (6), $\nabla_{\Gamma(t)} u = \mathcal{P}_{\Gamma(t)} \nabla u$, and hence for example

$$(\boldsymbol{w} \cdot \nabla_{\Gamma(t)} u)(\boldsymbol{\zeta}) \approx \boldsymbol{w}^T(\boldsymbol{\zeta}) \mathcal{P}_{\Gamma(t)}(\boldsymbol{\zeta}) \boldsymbol{\omega}_{\nabla}^T(\boldsymbol{\zeta}, \Xi_{\boldsymbol{\zeta}}) u_{\Xi_{\boldsymbol{\zeta}}}$$

as in (21), leading to the weights for the $\boldsymbol{\zeta}$-row of $\widetilde{K}(t, \boldsymbol{w}, \Gamma(t))$ in (13). Note that for convection dominated flows this approximation cannot be used as it is because of the stability issues: dominated convection terms may lead to the non-positiveness of a given numerical scheme and in such a way cause the appearance of negative values and give rise to nonphysical oscillations in the numerical solution. In this article though we do not discuss this issue.

Construction of the RBF-FD approximation of the term $u \nabla_{\Gamma(t)} \cdot \boldsymbol{v}$ in (14) is done by (22) in the form

$$(u \nabla_{\Gamma(t)} \cdot \boldsymbol{v})(\boldsymbol{\zeta}) \approx \omega_{\boldsymbol{\zeta}} u(\boldsymbol{\zeta}), \quad \omega_{\boldsymbol{\zeta}} = \operatorname{trace}\left(\mathcal{P}_{\Gamma(t)}(\boldsymbol{\zeta}) \boldsymbol{\omega}_{\nabla}^T(\boldsymbol{\zeta}, \Xi_{\boldsymbol{\zeta}}) \boldsymbol{v}_{\Xi_{\boldsymbol{\zeta}}}\right).$$

Hence, $G(t, \Gamma(t))$ in (14) is a diagonal matrix with the numbers $\omega_{\boldsymbol{\zeta}}$ on the diagonal.

4 Numerical Results

Here we demonstrate the applicability of the proposed RBF-FD scheme. In the following subsections we validate the spatial convergence of our scheme by considering an example of a heat equation on a curve. In the next subsections we show that the scheme can be applied not only to the surface evolution in the normal directions but also in the tangential one. In the last example we apply the RBF-FD approximation to convection dominated problems to demonstrate that additional stabilization techniques are required in this case.

4.1 Example 1

In the first test case we will validate the performance of the scheme and measure its accuracy by comparing with a given analytical solution. We solve the following equation

$$\frac{\partial^* u(\boldsymbol{x}, t)}{\partial t} = D\Delta_{\Gamma(t)} u(\boldsymbol{x}, t) + g(\boldsymbol{x}, t) \qquad \text{on} \quad \Gamma(t), \tag{25}$$

where $\Gamma(t)$ is prescribed as the zero level set of the function

$$\phi(\boldsymbol{x}, t) = |\boldsymbol{x}| - 1.0 + \sin(4\,t)(|\boldsymbol{x}| - 0.5)(1.0 - |\boldsymbol{x}|). \tag{26}$$

As a domain we choose $\Omega = \{\boldsymbol{x} \in \mathbb{R}^2 : 0.5 \leq |\boldsymbol{x}| \leq 1.0\}$. The boundary of the domain $\partial\Omega$ is aligned with a curve from the family $\Gamma_r(t) = \{\boldsymbol{x} | \phi(\boldsymbol{x}, t) = r\}$. The analytical solution is chosen to be

$$u(\boldsymbol{x}, t) = e^{-t/|\boldsymbol{x}|^2} \frac{x_1}{|\boldsymbol{x}|}. \tag{27}$$

Since $\Gamma(t)$ is time-dependent, Eq. (25) transforms into

$$\partial_t u + \boldsymbol{v}_S \cdot \nabla u + V\frac{\partial u}{\partial \boldsymbol{n}} - VHu + u\nabla_\Gamma \cdot \boldsymbol{v}_S - \Delta_\Gamma u = g, \tag{28}$$

where H is the mean curvature of $\Gamma(t)$ and therefore $H = -1/|\boldsymbol{x}|$. Substituting $v_S = 0$ into (28) we get

$$\partial_t u + V\frac{\partial u}{\partial \boldsymbol{n}} - VHu - \Delta_\Gamma u = g. \tag{29}$$

The function $u(\boldsymbol{x}, t)$ from (27) solves

$$\partial_t u - \Delta_\Gamma u = 0.$$

Therefore, one finds that

$$g = V\frac{\partial u}{\partial \boldsymbol{n}} - VHu = Vu\left(\frac{2t}{|\boldsymbol{x}|^3} + \frac{1}{|\boldsymbol{x}|}\right).$$

As the initial condition we set $u_{\text{init}} = u(\boldsymbol{x}, t = 0)$. Here, we calculate numerical solutions by the implicit scheme, $\theta = 1$ in (10), and the Crank-Nicolson schemes, $\theta = 1/2$ in (10). The corresponding mesh, as well as initial condition and analytical and numerical solutions are shown in Fig. 3a–d. Starting from $t = 0$, we calculate until the time point $T = 0.1$ with the time step $\Delta t \approx h^2$ by the Implicit-Euler scheme, $\theta = 1$ in (15), and $\Delta t \approx h$ by the second-order Crank-Nicolson scheme,

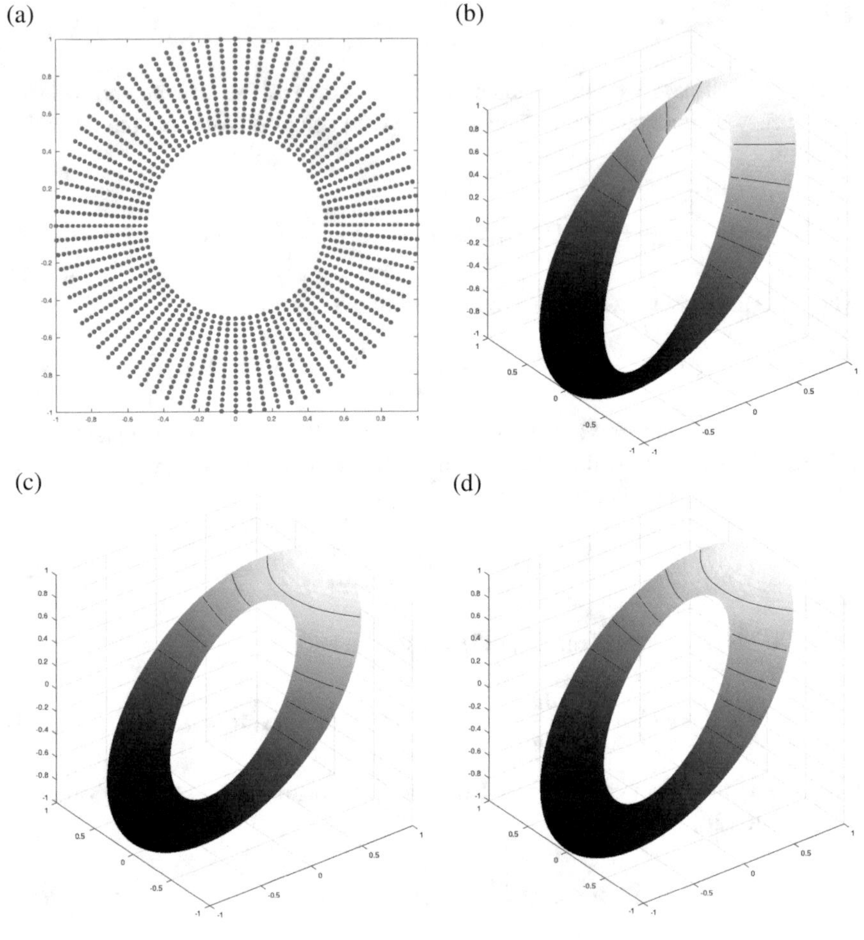

Fig. 3 Mesh, initial, analytical and numerical solutions. (**a**) Mesh, lev = 3. (**b**) Initial solution, lev = 4. (**c**) Analytical solution, lev = 4. (**d**) Numerical solution, lev = 4

$\theta = 1/2$ in (15). In Table 1 we measure the difference between the analytical and numerical solutions and obtain orders of convergence for the Implicit-Euler and Crank-Nicolson schemes. The corresponding error is defined as (cf. [6])

$$l_2(\Omega_\varepsilon)\text{-error} = \left(\frac{1}{|d.o.f.|} \sum_{\boldsymbol{x}_i \in \Omega_\varepsilon} \left| u_{\text{analyt}}(\boldsymbol{x}_i, T) - u_{\text{num}}(\boldsymbol{x}_i, T) \right|^2 \right)^{\frac{1}{2}},$$

One observes that the Crank-Nicolson scheme requires much fewer time steps in order to reach accuracy of the second order as the Implicit-Euler scheme.

Table 1 Convergence of the Implicit-Euler and Crank-Nicolson schemes

Lev.	d.o.f	Num. of time steps	$l_2(\Omega_{0.25})$-error	Order
Implicit scheme, $\Delta t \approx h^2$				
1	30	3	0.035854	–
2	100	10	0.009567	1.905
3	360	40	0.002602	1.878
4	1360	160	0.000748	1.798
5	5280	640	0.000213	1.812
Crank-Nicolson, $\Delta t \approx h$				
1	30	5	0.040218	–
2	100	10	0.09203	2.127
3	360	20	0.002367	1.959
4	1360	40	0.000673	1.814
5	5280	80	0.000192	1.809

4.2 Example 2

As our second test case we take Example 2 from [6]: we solve Eq. (25) in the domain $\Omega = \{\boldsymbol{x} \in \mathbb{R}^2 : 0.5 \leq |\boldsymbol{x}| \leq 1.0\}$ on the stationary level sets of

$$\phi(\boldsymbol{x}, t) = |\boldsymbol{x}| - 0.75.$$

Here, the initial solution is $u_0(\boldsymbol{x}) = \sin(4\gamma)$, where $\gamma \in [0, 2\pi)$ is the polar angle and the tangential velocity of the surface $\Gamma = \{\boldsymbol{x}|\phi(\boldsymbol{x}, t) = 0\}$ is $\boldsymbol{v}_S = 0$. Since $\gamma_t = 0$, the normal component of the surface velocity V is also zero. The mean value of u_0 vanishes on every level set Γ_r, hence the solution tends to zero as time tends to infinity. But this occurs at a rate which depends on the radius of the circle because of the different diffusion coefficients on the different circles. Numerical solutions at successive time instances are presented in Fig. 4. Here, the Implicit-Euler and Crank-Nicolson schemes deliver the same numerical results, with the difference that the Crank-Nicolson scheme requires much fewer time steps.

4.3 Example 3

In this test case we keep everything similar to the previous example in Sect. 4.2, but the tangential velocity of the surface is defined as

$$\boldsymbol{v}_S = 10\frac{\left(-\phi_{x_2}, \phi_{x_1}\right)}{|\nabla\phi|}. \tag{30}$$

Numerical results at some instances of time intervals are shown in Fig. 5. This example demonstrates that our approach is able to treat PDEs on time-dependent surfaces which move not only in the normal, but also in the tangential direction.

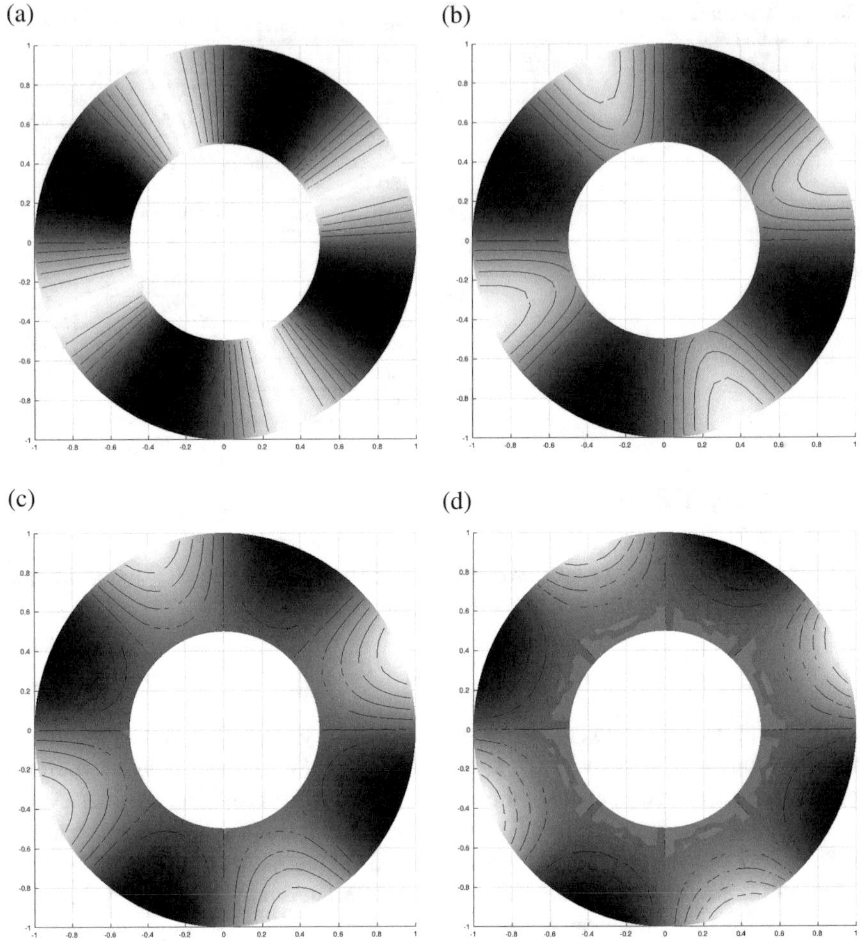

Fig. 4 Solution at various time instances, $\Delta t = 0.0001$. (**a**) Initial solution. (**b**) At $t = 0.02$. (**c**) At $t = 0.05$. (**d**) At $t = 0.1$

4.4 Example 4

In this example we consider a case, in which the evolution of a surface is with purely normal velocity:

$$\phi(\boldsymbol{x}, t) = |\boldsymbol{x}| - 0.75 + \sin(8\gamma)\,\sin(4t)\,(\boldsymbol{r} - 0.5)\,(1 - \boldsymbol{r}).$$

The initial solution is $u_0(\boldsymbol{x}) = \sin(8\gamma)$. Values of u in the boundary nodes are treated in the same way as in the inner ones. Figure 6a–c show evolutions of the obtained solution and of the level set function at different time instances. Here, we

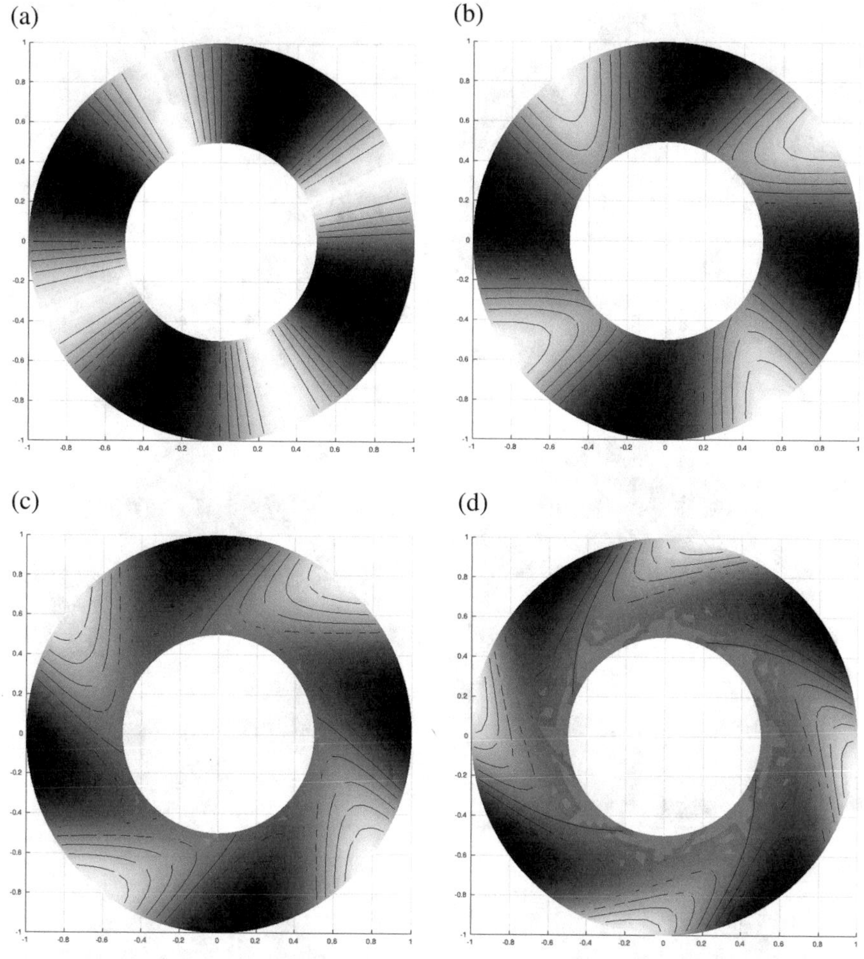

Fig. 5 Solution at various time instances, $\Delta t = 0.0001$. (**a**) Initial solution. (**b**) At $t = 0.002$. (**c**) At $t = 0.05$. (**d**) At $t = 0.1$

use a mesh at the sixth level of refinement (which corresponds to 5280 d.o.f). The time-step is $\Delta t = 0.0001$. For this kind of numerical simulations a number of stencil points and choice of their selection play an important role. For elliptic problems in 2D domains the corresponding study was done in works of Davydov et al. [3, 18]. For surface PDEs the relevant study is of a significant demand and is the aim of our future work.

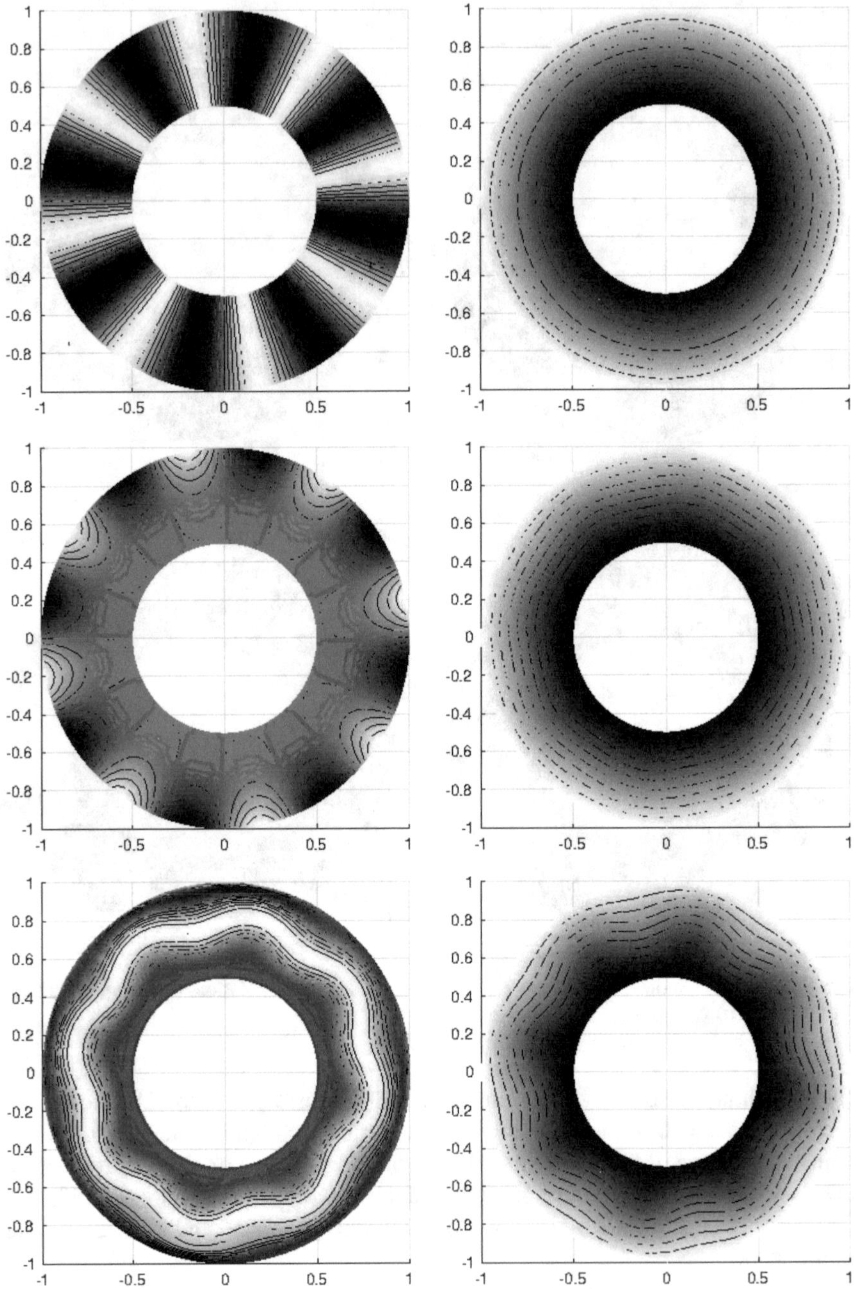

Fig. 6 Evolution of the numerical solution (left) and the level set function (right)

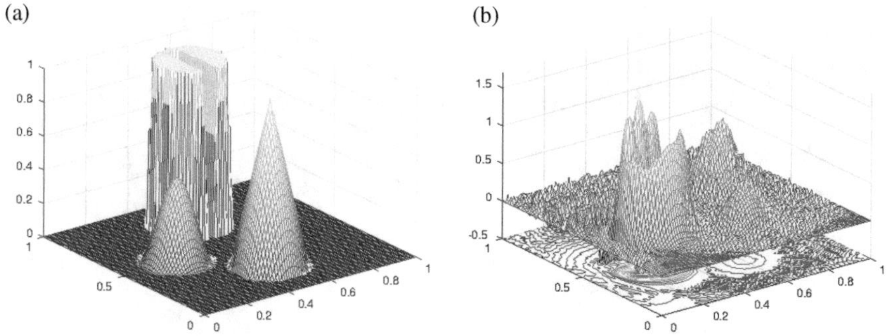

Fig. 7 The pure RBF-FD scheme for the transport problem. (**a**) Initial solution. (**b**) Numerical solution, $t = 1.0$

4.5 Example 5

The RBF-FD approximations of operator make it possible to perform numerical simulations for the pure transport equation:

$$\partial_t u + \boldsymbol{v} \cdot \nabla u = 0, \quad \text{in } \Omega = [0, 1]^2, \tag{31}$$

where $\boldsymbol{v} = (-y, x)$. As a simulation setting we choose $\Delta t = 0.001$ and $T = 3.0$. We place 6561 nodes in a Cartesian equidistant way inside Ω, which corresponds to the 81-by-81 refinement of the unit square. Initial conditions are taken from the solid-body rotation benchmark [14–16] and are shown in Fig. 7a.

The pure RBF-FD discretization for the transport problem does not guarantee mass conservation and does not keep numerical solution nonnegative. As a result, the nonphysical negative values of the numerical solution grow rapidly as time evolves, which might lead to an abnormal termination of the simulation run. In Fig. 7b we demonstrate the corresponding numerical solution at the time instance $t = 1.0$.

One can try to avoid this problem in many ways: to add nodes into those parts of the domain, where gradients of the numerical solution are large, to make the time-step smaller, to add some hyperviscosity into the model, or to use other stabilization techniques for the RBF-FD scheme. In Fig. 8b–d the numerical results which are obtained by using FCT techniques from works of Kuzmin et al. [14–16] are presented. The discussion about the applicability of the FCT methods for RBF-FD schemes remain out of the scope of this work and will be discussed elsewhere.

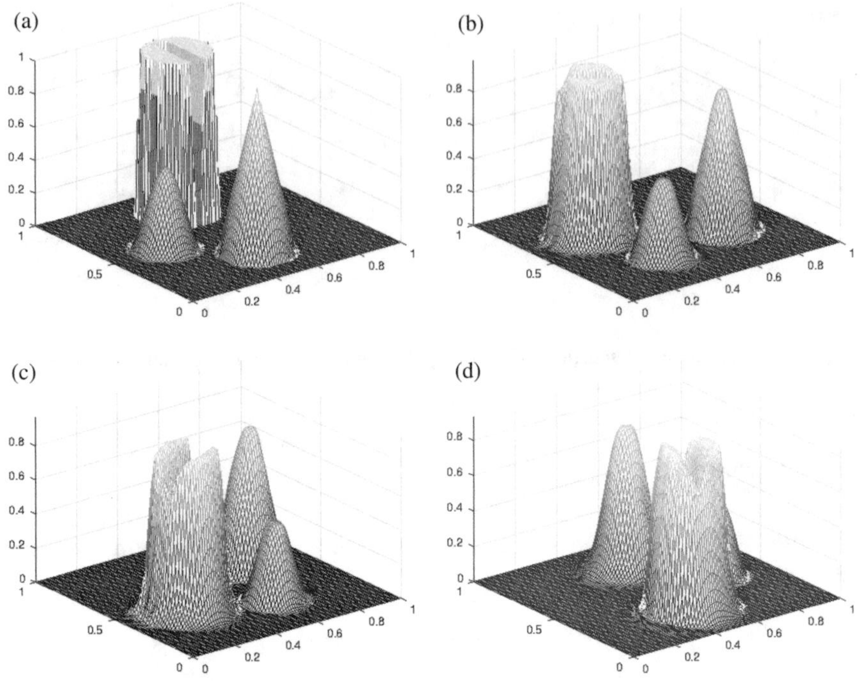

Fig. 8 The FCT stabilization technique for the RBF-FD scheme of the transport problem. (**a**) Initial solution. (**b**) Numerical solution, $t = 1.0$. (**c**) Numerical solution, $t = 2.0$. (**d**) Numerical solution, $t = 3.0$

5 Conclusion

In the current article we presented some methodology that allows the extension of the Radial Basis Function (RBF)-Finite Difference (FD) scheme to the numerical solution of partial differential equations (PDEs) of the reaction-diffusion type on an evolving-in-time hypersurface $\Gamma(t)$. Our numerical results confirm the reliability of the proposed computational framework in terms of numerical convergence and capturing of typical/expected solution profiles. We have thus developed an RBF-FD approach that can be employed for practical applications that involve PDEs on evolving surfaces.

The framework has a straightforward extension to three dimensional models which is mandatory when considering real-life applications, though some computational and code optimization are required, since the computational and analytical complexity significantly increases in three dimensional case. Detailed numerical investigations are subject of forthcoming work.

We also demonstrated that for convection dominated problems additional implementation of some stabilization technique is required to guarantee positivity preservation and non-oscillatory behavior of a numerical solution. In a follow-up paper we demonstrate that it is possible to efficiently adapt the Flux-corrected transport (FCT) technique to the proposed RBF-FD numerical scheme.

References

1. G.A. Barnett, N. Flyer, L.J. Wicker, Numerical methods for high dimensional Hamilton-Jacobi equations using radial basis functions. J. Comput. Phys. **196**(1), 327–347 (2004)
2. G.A. Barnett, N. Flyer, L.J. Wicker, An RBF-FD polynomial method based on polyharmonic splines for the Navier-Stokes equations: comparisons on different node layouts (2015). arXiv:1509.02615
3. O. Davydov, D.T. Oanh, Adaptive meshless centres and RBF stencils for Poisson equation. J. Comput. Phys. 230, 287–304 (2011)
4. O. Davydov, D.T. Oanh, On optimal shape parameter for Gaussian RBF-FD approximation of Poisson equation. Comput. Math. Appl. **62**, 2143–2161 (2011)
5. O. Davydov, R. Schaback, Error bounds for kernel-based numerical differentiation. Numer. Math. **132**, 243–269 (2016)
6. G. Dziuk, C.M. Elliott, An Eulerian approach to transport and diffusion on evolving implicit surfaces. Comput. Vis. Sci. **13**, 17–28 (2010)
7. G. Dziuk, C.M. Elliott, A fully discrete evolving surface finite element method. SIAM J. Numer. Anal. **50**(5), 2677–269 (2012)
8. G. Dziuk, C.M. Elliott, Finite element method for surface PDEs. Acta Numer. **50**(22), 289–396 (2013)
9. C.M. Elliott, B. Stinner, C. Venkataraman, Modelling cell motility and chemotaxis with evolving surface finite elements. J. R. Soc. Interface **9**(76), 3027–3044 (2012)
10. G.E. Fasshauer, Meshfree approximation methods with Matlab, in *Interdisciplinary Mathematical Sciences*, vol. 6 (World Scientific Publishers, Singapore, 2007)
11. N. Flyer, B. Fornberg, V. Bayona, G.A. Barnett, On the role of polynomials in RBF-FD approximations: I. Interpolation and accuracy. J. Comput. Phys. **321**, 21–38 (2016)
12. B. Fornberg, E. Larsson, N. Flyer, Stable computations with Gaussian radial basis functions. SIAM J. Sci. Comput. **33**, (2), 869–889 (2011)
13. E.J. Fuselier, G.B. Wright, A high-order kernel method for diffusion and reaction-diffusion equations on surfaces. J. Sci. Comput. **56**(3), 535–565 (2013)
14. D. Kuzmin, Explicit and implicit FEM-FCT algorithms with flux linearization. J. Comput. Phys. **228**, 2517–2534 (2009)
15. D. Kuzmin, M. Möller, Algebraic flux correction I. Scalar conservation laws, in *Flux-Corrected Transport: Principles, Algorithms and Applications* (Springer, Berlin, 2005), pp. 155–206
16. D. Kuzmin, S. Turek, Flux correction tools for finite elements. J. Comput. Phys. **175**, 525–558 (2002)
17. E. Larsson, E. Lehto, A. Heryudono, B. Fornberg, Stable computation of differentiation matrices and scattered node stencils based on Gaussian radial basis functions. SIAM J. Sci. Comput. **35**(4), A2096–A2119 (2013)
18. D.T. Oanh, O. Davydov, H.X. Phu, Adaptive RBF-FD method for elliptic problems with point singularities in 2D. Appl. Math. Comput. **313**, 474–497 (2017)
19. M.A. Olshanskii, A. Reusken, J. Grande, A finite element method for elliptic equations on surfaces. SIAM J. Numer. Anal. **47**(5), 3339–335 (2009)
20. M.A. Olshanskii, A. Reusken, X. Xu, An Eulerian space-time finite element method for diffusion problems on evolving surfaces. SIAM J. Numer. Anal. **52**(3), 1354–1377 (2014)
21. A. Rätz, A. Voigt, PDE's on surfaces — a diffuse interface approach. Commun. Math. Sci. **4**(3), 575–590 (2006)

22. V. Shankar, G.B. Wright, R.M. Kirby, A.L. Fogelson, A radial basis function (RBF)-Finite Difference (FD) method for diffusion and reaction-diffusion equations on surfaces. J. Sci. Comput. **63**(3), 745–768 (2015)
23. A. Sokolov, R. Ali, S. Turek, An AFC-stabilized implicit finite element method for partial differential equations on evolving-in-time surfaces. J. Comput. Appl. Math. **289**, 101–115 (2006)

A Data-Driven Multiscale Theory for Modeling Damage and Fracture of Composite Materials

Modesar Shakoor, Jiaying Gao, Zeliang Liu, and Wing Kam Liu

Abstract The advent of advanced processing and manufacturing techniques has led to new material classes with complex microstructures across scales from nanometers to meters. In this paper, a data-driven computational framework for the analysis of these complex material systems is presented. A mechanistic concurrent multiscale method called Self-consistent Clustering Analysis (SCA) is developed for general inelastic heterogeneous material systems. The efficiency of SCA is achieved via data compression algorithms which group local microstructures into clusters during the training stage, thereby reducing required computational expense. Its accuracy is guaranteed by introducing a self-consistent method for solving the Lippmann–Schwinger integral equation in the prediction stage. The proposed framework is illustrated for a composite cutting process where fracture can be analyzed simultaneously at the microstructure and part scales.

1 Introduction

The analysis and design of new materials with improved efficiency and performance requires cutting edge process and material modeling theories. For instance, new lightweight vehicles are being developed using lighter material systems [8]. Conventional processing-structure-property-performance relationships must be reconsidered to account for the microstructural complexity of new advanced materials systems such as hierarchical materials [16]. In this aim, integrated computational materials engineering approaches relying on predictive multiscale modeling theories are being developed [2, 13, 17].

M. Shakoor · J. Gao · W. K. Liu (✉)
Department of Mechanical Engineering, Northwestern University, Evanston, IL, USA
e-mail: w-liu@northwestern.edu

Z. Liu
Livermore Software Technology Corporation (LSTC), Livermore, CA, USA

M. Griebel, M. A. Schweitzer (eds.), *Meshfree Methods for Partial Differential Equations IX*, Lecture Notes in Computational Science and Engineering 129,
https://doi.org/10.1007/978-3-030-15119-5_8

In this paper, a data-driven multiscale modeling theory is presented and applied to a problem involving a process-structure relationship. This relationship emerges from microstructure modeling using computational homogenization and reduced order modeling.

The first novelty of the proposed data-driven multiscale modeling theory is the use of the so-called Self-consistent Clustering Analysis (SCA) [10]. This method relies on the Fast Fourier Transform (FFT) based numerical method introduced in Ref. [15], which formulates conventional balance equations with periodic boundary conditions as a periodic Lippmann–Schwinger equation. The originality in SCA is that the Lippmann–Schwinger equation is solved using a clustered discretization. The voxel mesh Direct Numerical Simulation (DNS) model of the microstructure is hence reduced into clusters of voxels, and degrees of freedom in the reduced model are defined cluster-wise instead of voxel-wise.

Voxels clustering is performed using the k-means clustering method [12] applied on a database of DNS results for the studied microstructure. These DNS results do not need to include complex loading paths, as accurate predictions could be obtained in a previous work using only proportional loading paths in 6 orthogonal directions [10]. In fact, in this previous work DNS results were obtained using small strain amplitudes for which material response remained in the linear elastic range.

The second novelty of the proposed data-driven multiscale modeling theory is its capability to model damage and fracture at multiple scales [11]. A concurrent computational homogenization scheme is developed in order to solve any macroscale problem with material laws computed on the fly from microstructure information. In this scheme, the macroscale problem is solved using the FE method, with the particularity that conventional phenomenological constitutive equations are replaced by micromechanical problems solved using SCA. These micromechanical models, called Representative Volume Elements (RVEs), include enough microstructural features to be statistically representative of the local microstructure around each material integration point.

Damage modeling leads to well-known localization and pathological mesh dependence issues. In the context of concurrent computational homogenization, these issues arise at two scales. Indeed, pathological localization can occur within arbitrary elements of the macroscale problem discretization, and also within arbitrary clusters of microscale problems discretizations. In the proposed data-driven multiscale damage modeling theory [11], the damage variable is regularized at the macroscale using non local integral averaging to avoid any localization between RVEs, while at the microscale it is coupled to constitutive equations in an average sense to avoid any localization within RVEs.

The paper is structured as follows. It starts with a presentation of SCA in Sect. 2, followed by details on its exploitation for multiscale damage modeling in Sect. 3. The relevance of the proposed data-driven multiscale modeling theory is illustrated by applications in Sect. 4.

A multiscale simulation of the cutting of a Unidirectional (UD) Carbon Fiber Reinforced Polymer (CFRP) composite is conducted to show how fracture can be modeled simultaneously at two scales with the proposed theory.

2 Self-consistent Clustering Analysis

In the Finite Element (FE) method, the displacement field is discretized at mesh nodes, and material integration is conducted at integration points. Reducing the number of displacement degrees of freedom does not directly reduce neither the number of integration points nor the cost of material integration. Therefore, FE based model order reduction methods must be coupled to material integration reduction techniques in order to be efficiently applicable to nonlinear materials [4–6].

In the FFT-based numerical method [15], the strain field is discretized voxel-wise, and material integration is conducted voxel-wise as well. As a consequence, reducing the number of strain degrees of freedom directly reduces the cost of both Lippmann–Schwinger equation solution and material integration. In comparison to FE based model order reduction methods [4–6], SCA is hence a more straightforward approach to reduced order modeling [10].

In the following, the superscript ^{m} indicates microscale variables that are discretized voxel-wise in the FFT-based numerical method, and cluster-wise in SCA. The RVE domain over which Eq. (2) is solved is denoted Ω^m. The superscript ^{M} indicates macroscopic variables that are homogeneous over the RVE.

2.1 Continuous Lippmann–Schwinger Equation

First order homogenization consists in defining the infinitesimal strain tensor field in the RVE ε^m as the addition of the macroscopic (homogeneous) strain ε^M and a microscopic (heterogeneous) fluctuation. As proved in Ref. [15], Hill's lemma enables to define the macroscopic Cauchy stress tensor σ^M as the average the microscopic one $\sigma^M = \frac{1}{|\Omega^m|}\int_{\Omega^m} \sigma^m(\mathrm{x})\mathrm{dx}$.

Hill's lemma requires $(\varepsilon^m - \varepsilon^M)$ to verify compatibility, i.e., to derive from a periodic displacement field, and σ^m to verify equilibrium, i.e. to be the solution of

$$\nabla.\sigma^m(\mathrm{x}) = 0, \mathrm{x} \in \Omega^m. \tag{1}$$

It can be shown that Eq. (1) is equivalent to

$$\varepsilon^m(\mathrm{x}) = -\int_{\Omega^m} \Phi^0(\mathrm{x}, \mathrm{x}') : \left(\sigma^m(\mathrm{x}') - \mathrm{C}^0 : \varepsilon^m(\mathrm{x}')\right) \mathrm{dx}' + \varepsilon^0, \mathrm{x} \in \Omega^m. \tag{2}$$

Equation (2) is the Lippmann–Schwinger equation for first order homogenization. The fourth rank tensor C^0 is the stiffness tensor associated to an isotropic linear elastic reference material. It will be determined in Sect. 2.2.2, as well as the far field strain tensor ε^0 and the periodic Green's operator Φ^0. The latter maps any tensor field τ^m to a compatible one:

$$\exists u \in (H^1(\Omega^m))^3,\ u \text{ periodic on } \Omega^m,\ -\Phi^0 * \tau^m = \frac{1}{2}(\nabla u + \nabla u^T). \tag{3}$$

The combination of Eqs. (2) and (3) yields a microscopic infinitesimal strain tensor ε^m that verifies compatibility and a Cauchy stress tensor σ^m that verifies equilibrium.

2.2 Discrete Lippmann–Schwinger Equation

SCA consists in solving Eq. (2) cluster-wise instead of voxel-wise. Figure 1a shows an example of voxel mesh for a single inclusion embedded within a matrix material. This voxel mesh is clustered in Fig. 1b. The clustering method for the training stage is presented in Sect. 2.2.1, including the construction of the database of DNS results. The use of this database to compute the mechanical response by solving the discrete Lippmann–Schwinger equation in the prediction stage is described in Sect. 2.2.2.

2.2.1 Training Stage

The aim of the training stage is to compute a cluster-wise discretization such as shown in Fig. 1. The mechanical response obtained by solving the Lippmann–Schwinger equation discretized cluster-wise should be as close as possible as that obtained by solving it voxel-wise. This can be done *a posteriori*, by solving the reduced order model for different trial configurations of clusters and searching for the optimal one. It can also be done a priori, for instance by basing the clustering algorithm on some mechanistic criterion such as the similarity in strain concentration tensors [10]. The strain concentration tensor field A^m is the fourth

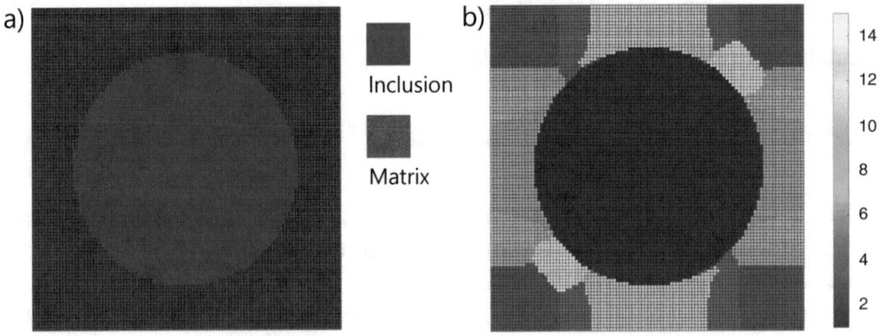

Fig. 1 Example of microstructure discretized using: (**a**) voxels; (**b**) clusters

order tensor field defined by

$$\varepsilon^m(\mathrm{x}) = A^m(\mathrm{x}) : \varepsilon^M, \mathrm{x} \in \Omega^m. \tag{4}$$

At a given instant T, the strain concentration tensor field depends on the applied macroscopic strain ε^M and, for plastic materials, on the loading history $\left(\varepsilon_t^M\right)_{t \leq T}$. It is neither possible to compute the A^m fields for all potential loading paths, nor is it possible to apply clustering directly to data of such high dimensionality. Therefore, the space of all possible A^m fields must be sampled down to a few loading paths [5]. The most cost-efficient way to do this is to consider only very small macroscopic strains ε^M in the training stage [10]. For such strains, the mechanical response is purely elastic and linear, and the single tensor field A^m, which has only 36 independent components due to symmetries of ε^m and ε^M, can be computed by conducting 6 DNS in 6 orthogonal loading directions.

The training data set hence consists in 36 values for each voxel of the DNS mesh. A k-means clustering algorithm [12] is applied to this data set. Since the microstructure is heterogeneous, clustering is done independently for each of its components, so that a given cluster cannot contain voxels from different components. The result of this training stage is a unique identifier $I = 1 \ldots k$ for each voxel of the DNS mesh. Voxels with same identifier have a similar microscopic response to macroscopic solicitations.

2.2.2 Prediction Stage

As a result of the training stage, the RVE domain Ω^m is discretized into k subsets $\left(\Omega_I^m\right)_{I=1\ldots k}$. The degrees of freedom in the FFT-based numerical method [15] are associated to the microscopic strain ε^m. In SCA [10], ε^m is discretized by a cluster-wise constant approximation $\left(\varepsilon_I^m\right)_{I=1\ldots k}$. As a consequence, the microscopic Cauchy stress tensor is also approximated cluster-wise $\left(\sigma_I^m\right)_{I=1\ldots k}$, and Eq. (2) can be discretized:

$$\varepsilon_I^m = - \sum_{J=1\ldots k} D_{IJ}^0 : \left(\sigma_J^m - \mathrm{C}^0 : \varepsilon_J^m\right) + \varepsilon^0, I = 1 \ldots k \tag{5}$$

where D^0 is the interaction tensor defined by

$$\begin{aligned} D_{IJ}^0 &= \frac{1}{|\Omega_I^m|} \int_{\Omega^m} \chi_I^m(\mathrm{x}) \int_{\Omega^m} \chi_J^m(\mathrm{x}')\Phi^0(\mathrm{x}, \mathrm{x}')\mathrm{dx}'\mathrm{dx} \\ &= \frac{1}{|\Omega_I^m|} \int_{\Omega_I^m} \left(\chi_J^m * \Phi^0\right)(\mathrm{x})\mathrm{dx}. \end{aligned} \tag{6}$$

The characteristic functions χ_I^m and χ_J^m are equal to 1 in, respectively, clusters I and J, and 0 elsewhere. In the FFT-based numerical method [15], the periodic Green's operator Φ^0 depends on C^0, and is known only in Fourier space. Because C^0 is associated to an isotropic linear elastic reference material, Φ^0 can be expressed in Fourier space as a function of the reference Lamé parameters λ^0 and μ^0. It is then obtained in real space by using the inverse FFT. In particular, Eq. (6) can be written in the form

$$D^0_{IJ} = f^1(\lambda^0, \mu^0) D^1_{IJ} + f^2(\lambda^0, \mu^0) D^2_{IJ},$$
$$D^i_{IJ} = \frac{1}{|\Omega^m_I|} \int_{\Omega^m_I} \text{FFT}^{-1} \left\{ \text{FFT}\{\chi^m_J\} \hat{\Phi}^i \right\} (\text{x}) \text{dx}, \, i = 1, 2. \tag{7}$$

The detailed expressions of f^1, f^2, $\hat{\Phi}^1$ and $\hat{\Phi}^2$ can be found in Refs. [9, 10, 15] among others. Drastic computational cost reduction is enabled by SCA thanks to a reduced number of degrees of freedom by clustering, and by the fact that D^1 and D^2 can be precomputed in the training stage. Therefore, neither FFTs no inverse FFTs are computed in the prediction stage, even if the reference material is changing.

In the present work, boundary conditions for Eq. (5) are purely kinematic. The average of the microscopic strain tensor ε^m must be enforced to be equal to the macroscopic strain tensor ε^M or, equivalently, the microscopic fluctuation must have zero average. This can be done by adding the condition $\sum_{I=1...k} |\Omega^m_I| \varepsilon^m_I = |\Omega^m| \varepsilon^M$ to Eq. (5).

As noted in Ref. [10], solutions of Eq. (5) are dependent on the choice of reference material. An optimal choice can be computed in the prediction stage by making the reference material consistent with the homogenized material. This means that the far field strain tensor ε^0 is an additional unknown that must be solved for in SCA [10], as opposed to the FFT-based numerical method where $\varepsilon^0 \equiv \varepsilon^M$ [15]. The self-consistent method consists in using a fixed-point iterative method where, at each step, the reference Lamé parameters λ^0 and μ^0 are changed so that $||\sigma^M - \text{C}^0 : \varepsilon^0||_2$ is minimized.

2.2.3 Summary

To summarize, the training stage in SCA consists in using a k-means clustering algorithm based on a mechanistic a priori clustering criterion computed using a simple sampling of the loading space. This training stage also includes computing all voxel-wise and computationally expensive operations such as FFTs and inverse FFTs.

In the prediction stage, a self-consistent iterative algorithm is used to search for the optimal choice of reference Lamé parameters. At each iteration of this self-consistent loop, matrix assembly operations are accelerated because all voxel-wise operations have been precomputed in the training stage and already reduced to cluster-wise contributions. A Newton-Raphson iterative algorithm must be embedded within each self-consistent iteration for nonlinear materials, in which case the discrete Lippmann–Schwinger equation is linearized.

The output from SCA are the microscopic variables' cluster-wise approximations, and the macroscopic Cauchy stress tensor.

3 Multiscale Damage

Concurrent computational homogenization implies introducing a macroscopic domain Ω^M, which can be a specimen or an industrial part. In the present work, the macroscale problem is solved using the FE method for the spatial discretization and an explicit scheme for the time discretization. The use of SCA as a material law is straightforward. Conventional constitutive equations defining the macroscopic stress σ^M as a function of the macroscopic strain ε^M are replaced by the theory described in Sect. 2. Although macroscopic variables are constant in space at the microscale, they vary at the macroscale, namely, $\sigma^M = \sigma^M(\mathrm{x}), \varepsilon^M = \varepsilon^M(\mathrm{x}), \mathrm{x} \in \Omega^M$.

If the relation between macroscopic variables σ^M and ε^M included a softening effect, then the macroscale problem would be ill-defined. Softening would localize in a single arbitrary layer of elements, which would be dependent on the FE mesh, and lead to zero dissipated energy for very fine meshes. This well-known pathological mesh dependence problem when modeling softening materials can be solved by using non local integral averaging on the macroscopic damage variable [1]. The main issue in concurrent computational homogenization is that there is no macroscopic damage variable, since damage is modeled within RVEs. While Hill's lemma allows to formulate σ^M as the average σ^m, there is no such result for internal variables related to plasticity or damage.

As proposed in a recent work [11], non local integral averaging can be applied directly on the microscopic damage variable d^m. An additional difficulty when damage is modeled within the RVE, is that the RVE problem itself becomes ill-defined if damage localizes within the RVE. To avoid such situation, damage can be uncoupled from the microscale problem, and considered only in an average sense. These two steps are detailed in the following.

3.1 Macroscale Damage

The damage variable d^m is defined at the microscale and discretized cluster-wise along with the infinitesimal strain tensor and the Cauchy stress tensor. While these microscale variables have been written as functions of microscale coordinates in Sect. 2, they must now be written as functions of both macroscale and microscale coordinates.

First, a RVE domain $\Omega^m = \Omega^m(\mathrm{x}^M)$ is associated to each point x^M of the macroscale domain Ω^M. Since the macroscale problem is solved using the FE method, RVEs are attached to the integration points of the macroscale FE mesh.

Second, microscale variables can be written as functions of both macroscale and microscale coordinates. For instance, the microscale damage variable is discretized as

$$d^m(\mathrm{x}^M, \mathrm{x}^m) = \sum_{I=1...k} d_I^m(\mathrm{x}^M)\chi_I^m(\mathrm{x}^M, \mathrm{x}^m), \mathrm{x}^M \in \Omega^M, \mathrm{x}^m \in \Omega^m(\mathrm{x}^M). \tag{8}$$

Third, classic non local integral regularization [1] can be applied to the microscale damage variable defined in Eq. (8), with the novelty that the averaging is applied at two scales [11]. The non local microscale damage variable $\overline{d}^m$ is hence defined by

$$\begin{aligned} \overline{d}^m(\mathrm{x}^M, \mathrm{x}^m) &= \sum_{I=1...k} \overline{d}_I^m(\mathrm{x}^M)\chi_I^m(\mathrm{x}^M, \mathrm{x}^m), \mathrm{x}^M \in \Omega^M, \mathrm{x}^m \in \Omega^m(\mathrm{x}^M), \\ \overline{d}_I^m(\mathrm{x}^M) &= \int_{\Omega^M} w(||\mathrm{x}^M - \mathrm{y}^M||_2) d_I^m(\mathrm{y}^M) \mathrm{dy}^M, \mathrm{x}^M \in \Omega^M, \end{aligned} \tag{9}$$

where w is the non local averaging kernel given by

$$\begin{aligned} w(r) &= \frac{w_\infty(r)}{\int_0^{+\infty} w_\infty(r')\mathrm{d}r'}, r \in [0, +\infty[, \\ w_\infty(r) &= \begin{cases} \left(1 - 4\frac{r^2}{l_c^2}\right)^2, & r \le l_c \\ 0, & r > l_c \end{cases}, r \in [0, +\infty[. \end{aligned} \tag{10}$$

The characteristic length scale l_c is a material parameter associated to the width of damage localization bands at the macroscale. As defined by Eqs. (9) and (10), the non local damage variable is regularized at the macroscale and the macroscale FE problem is hence well-defined. In particular, results will not pathologically depend on the macroscale FE mesh. However, the non local damage variable may still localize at the microscale and yield clustering-dependent results.

3.2 Microscale Damage

To prevent localization within RVEs, an averaging procedure is also applied at the microscale. This procedure consists in uncoupling the damage model from the plasticity model, and modeling softening only in an average sense.

First, the microscopic infinitesimal strain tensor ε^m is additively decomposed into an elastic part $\varepsilon^{m,el}$ and a plastic part $\varepsilon^{m,pl}$. The microscopic damage variable d^m is written as a function of the plastic strain $d^m = d^m(\varepsilon^{m,pl})$, but the plastic strain itself is not a function of the damage variable. Equation (5) is hence solved with a first definition of the microscopic Cauchy stress tensor that does not account for softening.

Second, the evolution of the damage variable is computed based on the stress state and plastic strain computed in the first step. For the CFRP composite studied in Sect. 4, the following power law is used to define the evolution of damage in the epoxy matrix as a function of the von Mises equivalent plastic strain $\varepsilon^{m,pl,eq}$:

$$d^{epoxy} = 1 - \frac{\varepsilon^{m,pl,c}}{\varepsilon^{m,pl,eq}} \exp\left(-100(\varepsilon^{m,pl,eq} - \varepsilon^{m,pl,c})\right) \tag{11}$$

This law involves the material parameter $\varepsilon^{m,pl,c} = 0.13$. The brittle fracture of fibers is modeled by maximum stress theory [3]. Thus, the damage variable in fibers can be equal only to 0 or 1.

Third, the effective macroscopic Cauchy stress tensor σ^M is computed by solving Eq. (5) with a softening effect but no plasticity, namely, C^m being the microscopic elastic stiffness tensor, $\sigma^m = (1 - \overline{d}^m)C^m : \varepsilon^{m,el}$. The applied macroscopic strain for this third step is the macroscopic elastic strain computed by elastic relaxation and averaging of the first step solution [11].

Although this averaging procedure requires two solutions of Eq. (5), only the first one accounts for plasticity and material nonlinearity. Thus, the second solution has a reduced cost. Furthermore, both solutions are accelerated thanks to SCA.

Because the microscopic plastic strain does not depend on damage, it can not localize pathologically within a single layer of clusters. Then, the damage variable being written as a function of the microscopic plastic strain, pathological localization of this variable is not possible within the RVE. With the addition of the macroscale non local integral averaging described in Sect. 3.1 that prevents pathological localization and mesh dependence at the macroscale scheme, a regularized multiscale damage theory is obtained.

4 Multiscale Carbon Fiber Reinforced Polymer Composite Cutting Process Modeling

An example of concurrent simulation of the cutting process of a UD CFRP composite is proposed in this section. Literature reports some progress made in simulating CFRP cutting processes at the microscale [3, 7]. This is necessary to observe microscale deformation, such as fiber distortion and matrix cracking during this process. For a full scale cutting process, the material is generally assumed homogeneous and modeled using phenomenological constitutive equations to reduce the computational cost, but all microscale details are lost. The theory presented in this paper opens a new window for structure scale simulation with minimum loss of microscale details.

The cutting simulation is based on experimental work performed in a previous study [3]. Details of the experimental setup can be found in the given reference. To demonstrate the capabilities of the multiscale modeling theory presented in this paper, a 3D transverse UD CFRP cutting simulation is performed on a domain of

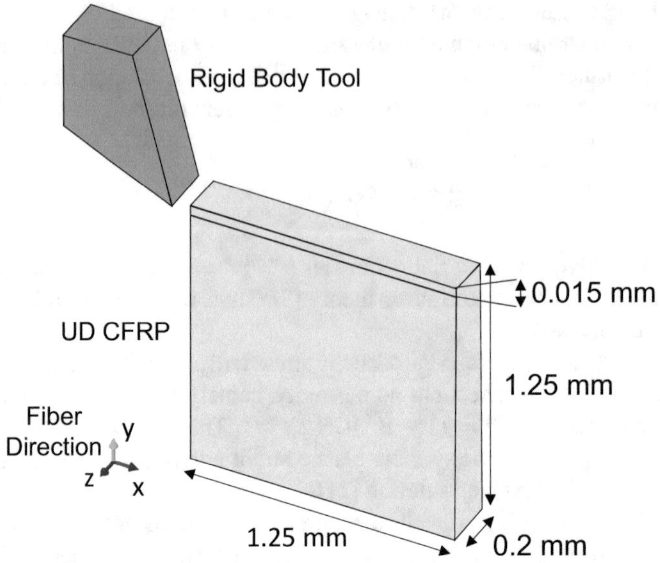

Fig. 2 Cutting simulation setup and geometry of UD CFRP part

length × weight × height of 1.25 × 0.2 × 1.25 mm. The cutting depth is 0.015 mm, following the experimental setup. The model setup is shown in Fig. 2. The bottom surface of the UD CFRP part is fixed to ensure that it stays in its position. Cutting speed is set to 8 mm/s according to the experimental setup. The UD CFRP part is modeled with 142,500 reduced integration cubic elements, where each element has one integration point.

The UD CFRP material has fiber volume fraction of 60%. Fiber is assumed to be of circular shape with a diameter of 7 μm, as shown in Fig. 3a. The RVE has an identical length and width of 84 μm, and a depth of 14 μm. The RVE is meshed

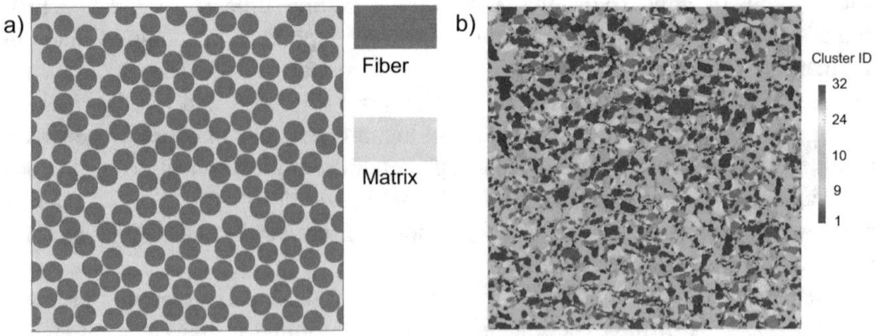

Fig. 3 Cross section of the UD CFRP RVE showing: (**a**) the random fibers arrangement; (**b**) the clusters

Table 1 Carbon fiber and epoxy matrix elastic properties

E_1	E_2	E_3	υ_{12}	υ_{13}	υ_{23}	E_m	υ_m
240 GPa	19 GPa	19 GPa	0.28	0.28	0.32	3.8 GPa	0.387

with 740 × 740 × 5 voxels, which are then clustered using the method presented in Sect. 2.2.1 into 16 clusters for the matrix, and 16 additional clusters for the fibers. The result is shown in Fig. 3b.

Fiber and matrix elastic properties are given in Table 1. Fibers tensile and compressive strengths follow the parameters listed in Ref. [3]. It is assumed that excessive deformation of the matrix happens when the cutting tool is in compressive contact with the material. Thus, matrix plasticity has been calibrated to the uniaxial compression curve for epoxy in Fig. 1 of Ref. [14] with a simple J_2 plasticity model. The damage evolution law for the matrix has been given in Eq. (11).

Using the damage evolution law in Eq. (11) within the multiscale damage modeling theory, the microscale damage variable might reach 1 in some clusters. In such case, the material has completely lost its load carrying capacity. If this happens for multiple clusters, the averaged load carrying capacity of some RVEs might be significantly lost. With a criterion to measure this loss of averaged load carrying capacity, element deletion could be triggered in the macroscale mesh to model the cutting process. A macroscopic non local damage variable $\overline{d}^M$ is introduced to measure this loss of averaged load carrying capacity:

$$\overline{d}^M = 1 - \frac{||\sigma^M : \sigma^M_{pl}||}{||\sigma^M_{pl} : \sigma^M_{pl}||} \tag{12}$$

where σ^M_{pl} is the average of the Cauchy stress tensor computed with the plasticity model but no damage, while σ^M is the macroscopic Cauchy stress tensor computed with the non local damage model but no plasticity. For each element of the UD CFRP part FE model, element deletion is triggered when $\overline{d}^M = 0.25$.

Simulation of concurrent UD CFRP cutting has been performed for 0.01 s using ABAQUS CAE with the multiscale damage model. The average reaction force obtained from concurrent cutting was 0.881 N. The comparison between numerical result and experimental result is presented in Table 2. The simulated average horizontal cutting force is 7.3% less than that measured in the experiment.

The main feature of the multiscale model is that it captures the microscale fiber and matrix failure within UD RVEs at each integration point of the macroscale part. Figure 4 shows the macroscopic part with RVEs at three selected integration points where damage can be seen at different phases of the cutting process. At different

Table 2 Comparison of simulated cutting force against experimental data

	Experimental data [3]	Multiscale model	Difference
Horizontal cutting force	0.946 N/m	0.881 N/m	7.3%

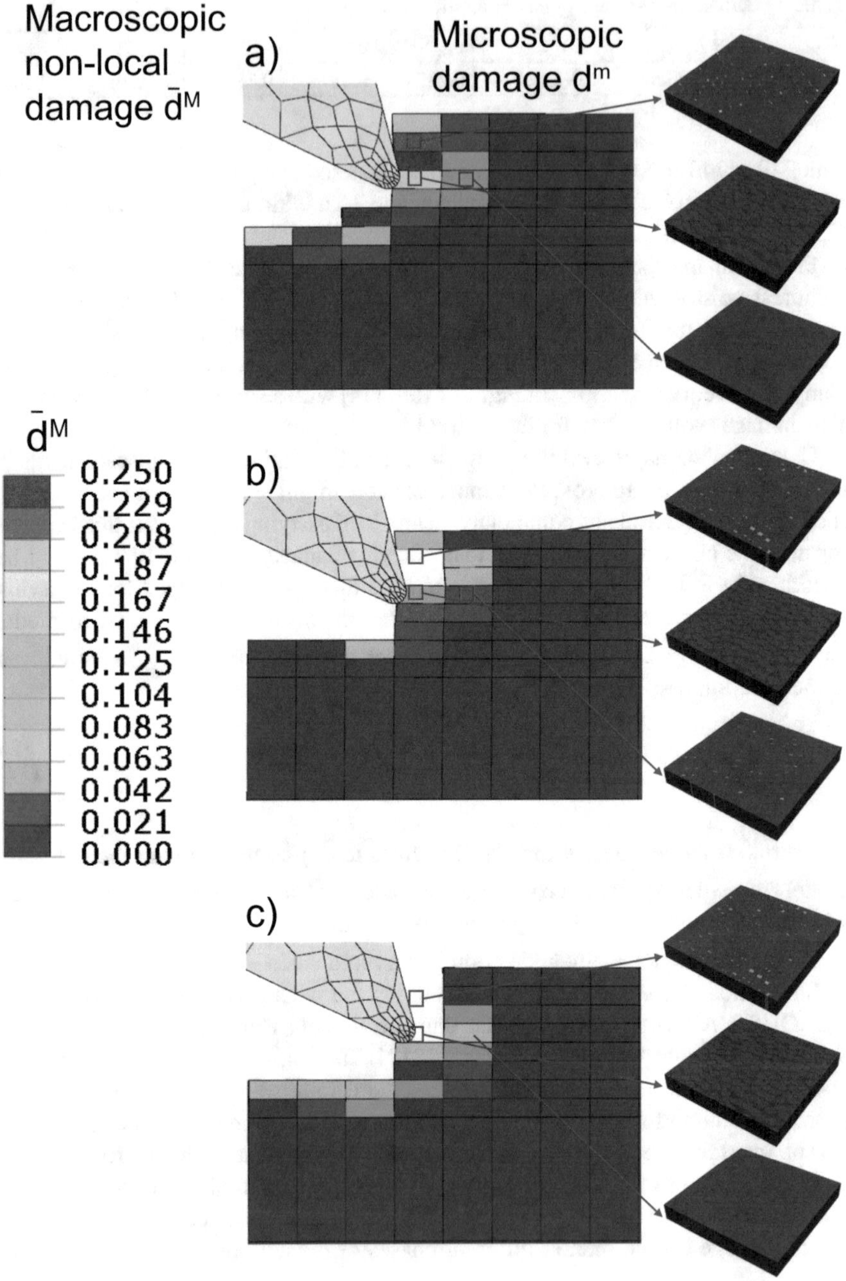

Fig. 4 Macroscopic non local damage variable and microscopic damage variable at: (**a**) 5.125e−3 s; (**b**) 5.250e−3 s; (**c**) 5.5e−3 s

time steps, it can be seen that different elements have different macroscopic non local damage $\overline{d}^M$ that can be traced back to the microscopic damage d^m within each cluster in RVEs. Here, the element embedding the second RVE fails after the element embedding the first RVE, although it seems to have a higher total damaged volume. This shows the effect of the element deletion criterion in Eq. (12), which does not define the macroscopic damage variable just as the average of the microscopic one, but as the actual loss of load carrying capacity. Additionally, the local damage within damaged elements RVEs is transferred to neighboring elements RVEs via non local averaging. This can be seen from the left column of Fig. 4, where localized damage is being distributed to nearby elements from those contacting the tool. As a consequence, some damage can be seen in the third RVE, but it does not cause enough loss of load carrying capacity for the associated element to get deleted.

5 Conclusions

Two main contributions were presented in this paper. A regularized multiscale modeling theory was proposed to model multiscale damage and fracture processes such as the fracture of material systems with heterogeneous microstructure. The latter was modeled using the self-consistent clustering analysis method for data-driven reduced order modeling.

To illustrate the capabilities of this data-driven multiscale damage and fracture modeling theory, a simulation of a cutting process was conducted. The considered material, a carbon fiber reinforced composite, exhibited a heterogeneous microstructure which failed by epoxy matrix damage and fiber breakage. The effect of these microscale damage and fracture mechanisms on the macroscale behavior was modeled using a material law computed on-the-fly by the multiscale modeling theory instead of relying on conventional phenomenological constitutive equations.

For future work, the proposed theory is going to be extended to other material and processes involving more complex damage and fracture mechanisms.

Acknowledgements Modesar Shakoor and Wing Kam Liu warmly thank the support from National Institute of Standards and Technology and Center for Hierarchical Materials Design (CHiMaD) under grant No. 70NANB13Hl94 and 70NANB14H012. Jiaying Gao and Wing Kam Liu warmly thank the support from DOECF-ICME project under grant No. DE-EE0006867. Zeliang Liu would like to thank Dr. John O. Hallquist of LSTC for his support.

References

1. Z.P. Bažant, M. Jirásek, Nonlocal integral formulations of plasticity and damage: survey of progress. J. Eng. Mech. **128**(11), 1119–1149 (2002)
2. M.A. Bessa, R. Bostanabad, Z. Liu, A. Hu, D.W. Apley, C. Brinson, W. Chen, W.K. Liu, A framework for data-driven analysis of materials under uncertainty: countering the curse of dimensionality. Comput. Methods Appl. Mech. Eng. **320**, 633–667 (2017)

3. H. Cheng, J. Gao, O.L. Kafka, K. Zhang, B. Luo, W.K. Liu, A micro-scale cutting model for ud cfrp composites with thermo-mechanical coupling. Compos. Sci. Technol. **153**, 18–31 (2017)
4. F. Chinesta, A. Leygue, F. Bordeu, J.V. Aguado, E. Cueto, D. Gonzalez, I. Alfaro, A. Ammar, A. Huerta, PGD-based computational vademecum for efficient design, optimization and control. Arch. Comput. Methods Eng. **20**(1), 31–59 (2013)
5. O. Goury, D. Amsallem, S.P.A. Bordas, W.K. Liu, P. Kerfriden, Automatised selection of load paths to construct reduced-order models in computational damage micromechanics: from dissipation-driven random selection to Bayesian optimization. Comput. Mech. **58**(2), 213–234 (2016)
6. J.A. Hernández, J. Oliver, A.E. Huespe, M.A. Caicedo, J.C. Cante, High-performance model reduction techniques in computational multiscale homogenization. Comput. Methods Appl. Mech. Eng. **276**, 149–189 (2014)
7. D. Iliescu, D. Gehin, I. Iordanoff, F. Girot, M.E. Gutiérrez, A discrete element method for the simulation of CFRP cutting. Compos. Sci. Technol. **70**(1), 73–80 (2010)
8. W.J. Joost, Reducing vehicle weight and improving U.S. energy efficiency using integrated computational materials engineering. JOM **64**(9), 1032–1038 (2012)
9. M. Kabel, T. Böhlke, M. Schneider, Efficient fixed point and Newton–Krylov solvers for FFT-based homogenization of elasticity at large deformations. Comput. Mech. **54**(6), 1497–1514 (2014)
10. Z. Liu, M.A. Bessa, W.K. Liu, Self-consistent clustering analysis: an efficient multi-scale scheme for inelastic heterogeneous materials. Comput. Methods Appl. Mech. Eng. **306**, 319–341 (2016)
11. Z. Liu, M. Fleming, W.K. Liu, Microstructural material database for self-consistent clustering analysis of elastoplastic strain softening materials. Comput. Methods Appl. Mech. Eng. **330**, 547–577 (2018)
12. J. Macqueen, Some methods for classification and analysis of multivariate observations, in *Proceedings of the Fifth Berkeley Symposium on Mathematical Statistics and Probability*, vol. 1 (University of California Press, Berkeley, 1967), pp. 281–297
13. K. Matouš, M.G.D. Geers, V.G. Kouznetsova, A. Gillman, A review of predictive nonlinear theories for multiscale modeling of heterogeneous materials. J. Comput. Phys. **330**, 192–220 (2017)
14. A.R. Melro, P.P. Camanho, F.M.A. Pires, S.T. Pinho, Micromechanical analysis of polymer composites reinforced by unidirectional fibres: Part I–constitutive modelling. Int. J. Solids Struct. **50**(11), 1897–1905 (2013)
15. H. Moulinec, P. Suquet, A numerical method for computing the overall response of nonlinear composites with complex microstructure. Comput. Methods Appl. Mech. Eng. **157**(1–2), 69–94 (1998)
16. J.H. Panchal, S.R. Kalidindi, D.L. McDowell, Key computational modeling issues in integrated computational materials engineering. Comput. Aided Des. **45**(1), 4–25 (2013)
17. J. Smith, W. Xiong, W. Yan, S. Lin, P. Cheng, O.L. Kafka, G.J. Wagner, J. Cao, W.K. Liu, Linking process, structure, property, and performance for metal-based additive manufacturing: computational approaches with experimental support. Comput. Mech. **57**(4), 583–610 (2016)

Modeling the Friction Drilling Process Using a Thermo-Mechanical Coupled Smoothed Particle Galerkin Method

Cheng-Tang Wu, Youcai Wu, Wei Hu, and Xiaofei Pan

Abstract This paper presents an up-to-date Lagrangian particle method for the analysis of a coupled thermo-mechanical problem in the friction drilling simulation. The method is obtained by a modification of variational equations using the penalized approach to avoid onerous stability problems in conventional Lagrangian particle methods and to obtain semi-discrete equations that are amenable to temporal and spatial integration using the staggered explicit time marching scheme. To deal with the critical numerical limitation associated with large deformation and material separation at the bushing forming stage, the method is furnished with an adaptive anisotropic Lagrangian kernel and a bond-based failure criterion. Representative simulation of a thermal-mechanical coupled friction drilling process is studied, and results are compared with the experimental data to examine the validity of this study.

1 Introduction

Friction drilling is a nonconventional drilling process that utilizes the heat generated by friction between the rotating tool and metal workpiece to soften the material and create a hole [1]. Unlike traditional drilling, friction drilling is a chip-less and dry manufacturing method that produces the hole in only one operation without the material removal and lubricants. Friction drilling creates sturdy bushing on thin walled structures such as sheet metal or tubing. The bushing created in the process is usually two to three times thicker than the original workpiece allowing for mounting of soldered and screw connections in a simple and efficient way. Friction drilling can be performed on most metal materials using a high-speed rotating tool made of conical tungsten carbide. Typical applications of friction drilling in automotive industry include seat handle/frame, foot pedal, exhaust part, fuel rail, and among

C.-T. Wu (✉) · Y. Wu · W. Hu · X. Pan
Livermore Software Technology Corporation, Livermore, CA, USA
e-mail: ctwu@lstc.com

M. Griebel, M. A. Schweitzer (eds.), *Meshfree Methods for Partial Differential Equations IX*, Lecture Notes in Computational Science and Engineering 129,
https://doi.org/10.1007/978-3-030-15119-5_9

others. A growing interest on the study of friction drilling process has been shown by many car companies motivated by the need to reduce manufacturing costs and obtain the high quality of final product.

Numerical modeling is a necessary tool to understand the material flow, temperatures, stresses and strains which are difficult to measure experimentally during friction drilling [2]. Numerical simulation of friction drilling involves solving a coupled thermo-mechanical system, a task that can turn out to be difficult when considerable deformation and material separation are developed in bushing forming. It has become one of the research topics of great interest in computational mechanics over the last years. Since the Eulerian representation of a material has the difficulty to capture the free surface flow in the simulation of bushing forming, Lagrangian finite element methods [3] have been favored. While the Lagrangian finite element method is used in combination with the r-adaptive re-meshing strategy [4, 5] to handle large deformation problems in similar manufacturing processes such as the friction stir spot welding (FSSW) and the friction stir welding (FSW) [6, 7], modeling material separation in the friction drilling process has always been problematic. This is because the r-adaptive re-meshing may become unstable or unable to maintain the high quality mesh when some or lots elements are deleted using the element erosion technique in mimicking the material separation phenomenon during the forming of the metal bushing.

In comparison to Lagrangian finite element methods, Lagrangian particle methods are adventurous in modeling large deformation and material failure [8–10] problems. Lagrangian particle methods were also found to be very effective on reducing volumetric locking and shear locking in solid and structural analyses [11, 12]. Smoothed Particle Hydrodynamics (SPH) method developed by Gingold and Monaghan [13] and Lucy [14] in late 1970s for astrophysical problems has been considered the earliest Lagrangian particle method. In early 1990s, Libersky and Petschek [15] extended SPH to solid mechanics applications. In spite of its popularity in simulating high-velocity impact/penetration and fluid flow problems [16], SPH has limited success in solid mechanics applications due to several numerical instabilities. Among them, tensile instability [17], spurious zero-energy mode [18] and excessive straining [19] are critical to the simulation result and have been the important research topics in the past two decades.

Intensive research work has been carried out to resolve those numerical instabilities. For instance, the introduction of Lagrangian kernel [8, 20] or stress points method [21] has been proven to effectively remove the tension instability in Lagrangian particle methods. The origin of spurious zero-energy mode can be explained by inspecting the system of equations of the particle method. A pioneering approach to circumvent this numerical instability was demonstrated by Beissel and Belytschko [22] using a residual-type stabilization procedure. A variant of this stabilization approach includes the non-residual type of stabilization methods [23, 24], stabilized conforming nodal integration (SCNI) method [25], and variationally consistent integration methods [26]. The problem of excessive straining emerges as a numerical instability in Lagrangian particle methods when the strictly use of Lagrangian kernel is no more applicable in large deformation range. In order to enable the Lagrangian kernel in large deformation analyses, semi-Lagrangian kernel [27] and adaptive anisotropic Lagrangian kernel [28] have been developed. Nevertheless, very few studies [24, 29] have addressed all numerical issues concurrently and comprehensively.

Smoothed Particle Galerkin (SPG) method motivated by Beissel and Belytschko's residual-type stabilization method [22] is one of the new Lagrangian particle methods developed by Wu et al. [29] to deal with those numerical instabilities. Another new Lagrangian particle method which is based on implicit gradient expansion [30], strain gradient stabilization technique [25] and semi-Lagrangian kernel [27] was proposed by Hillman and Chen [24] to sufficiently control those numerical instabilities in severe deformation analysis. These two Lagrangian particle methods share a common feature in augmenting the standard quadratic energy functional by a non-residual term for stabilization. Since the stabilization in those methods is accomplished without the use of the momentum equation residual, dependence of artificial control parameters for stabilization can be eliminated.

Modeling material separation in three-dimensional problem is another important research topic for Lagrangian particle methods as well as a desirable feature for industrial applications. However, the extant literature in Lagrangian particle methods gives very few examples [9] in simulating the three-dimensional material separation process. In essence, the development of 3D material separation techniques for Lagrangian particle methods face formidable challenges in tracing moving discontinuity surfaces and in dealing with the interaction of particles affecting by the discontinuity. In order to avoid those numerical difficulties and meet the current need in industrial applications, a bond-based failure criterion inspired by the peridynamics method of Silling et al. [31] was introduced to SPG method by Wu et al. [29] for material failure analysis. While the SPG method has been used to model ductile failure in metals recently [32], its application to the coupled thermo-mechanical problem in manufacturing applications remains to be developed.

The object of this study is to develop a thermo-mechanical coupled SPG method to realistically simulate the friction drilling process involving large deformation and material separation. The reminder of the paper is organized as follows: the preliminaries and weak formulations for the coupled thermo-mechanical problem are given in Sect. 2. In Sect. 3, the SPG formulation and semi-discrete equations are provided. The computational procedures for thermal and mechanical induced large deformation and material separation analyses are described in the Sect. 4. One friction drilling simulation using the present method is given in Sect. 5, and conclusions are made in Sect. 6.

2 Preliminaries

The highly coupled and nonlinear system in thermo-mechanical equations for the friction drilling simulation is usually difficult to be solved by the simultaneous time-stepping algorithm. In particular, the large and un-symmetric system in fully coupled thermo-mechanical equations inevitably involves the convergent problem and is expensive to be solved implicitly in the presence of large deformation, material separation, severe contact conditions and contact-induced thermal shock. Additionally, because friction drilling is a very quick machining process, staggered and explicit time-stepping schemes are considered in this study for the application of interest. In the *staggered time-steeping algorithm* [33], the thermo-mechanical coupled system of equations is partitioned into a thermal phase at fixed configuration, followed by a mechanical phase at constant temperature.

In the thermal phase of the coupled system, we consider the transient heat transfer response of a metal workpiece in three-dimensional case. We assume linear dependence of heat flux on the temperature gradient which is known as the Fourier's law. We also assume isotropy of the material thermal conductivity. Since the temperature range over which the workpiece is observed in experiments is lower than the melting point, we presume the drilling process does not involve the material phase change. We also presume the heat generation is only due to plastic deformation and frictional contact between the drilling tool and workpiece. If we neglect the thermal exchange due to surface convection and radiation in the workpiece during the friction drilling, the standard variational formulation of the thermal energy conservation equation can be written to find the temperature field $\theta(X,t) \in \Theta = \{\theta \in H^1(\Omega) : \theta = \theta_d \text{ on } \partial\Omega_{\mathrm{d}}\}$ such that for arbitrary variation $\delta\theta \in \Theta_0 = \{\theta \in H^1(\Omega) : \theta = 0 \text{ on } \partial\Omega_d\}$ the following equation is satisfied

$$\int_{\Omega} \rho C_p \dot{\theta} \delta\theta \, d\Omega + \int_{\Omega} k \nabla\theta \cdot \nabla(\delta\theta) \, d\Omega = \int_{\partial\Omega_n} q_n \delta\theta \, ds + \int_{\Omega} Q \delta\theta \, d\Omega + \int_{\partial\Omega_c} h_c(\theta - \theta_{\mathrm{tool}}) \delta\theta \, ds + \int_{\partial\Omega_c} \eta\tau \cdot \left[\dot{u}^t\right] \delta\theta \, ds. \tag{1}$$

In the above equation ρ is the mass density, C_p is the heat capacity, k is the isotropic thermal conductivity, ∇ is the gradient operator with respect to current position x, and "$\nabla\cdot$" denotes the divergence operator. $\partial\Omega_d$ describes a Dirichlet boundary imposed by a temperature θ_d, and $\partial\Omega_n$ is the Neumann boundary prescribed by a normal heat flux $q_n = k(\theta)\nabla\theta \cdot n$, where n is the outward unit normal vector. We also have Q denoting the internal heat generation rate per unit deformed volume from plastic deformation and is defined by

$$Q := \eta S : \dot{\varepsilon}^p \tag{2}$$

where S and $\dot{\varepsilon}^p$ are the deviatoric part of Cauchy stress and the rate of plastic straining, respectively, and η is the Taylor-Quinney [34] coefficient that takes into account the fraction of heat generated by plastic deformation energy dissipation. The boundary $\partial\Omega_c$ denotes the contact surface with a thermal exchange between the tool and work piece. Subsequently, the third term on the right-hand side of Eq. (1) designates the interfacial heat transfer where h_c is the heat conductance on $\partial\Omega_c$, and

θ_{tool} is the temperature of the tool. The last term on the right-hand side of Eq. (1) represents the rate of frictional energy dissipation in which η is the fraction of heat generated by the frictional contact, and τ is the Cauchy contact traction and $\left[\dot{u}^t\right]$ is the contact slip rate which is regarded as the jump in velocity across contact surface.

In the mechanical phase, the dynamic process of friction drilling process is described by the equation of motion in the context of large strain analysis. During the friction drilling process, the workpiece experiences different rates of heating and cooling, and thus expansion and contraction. This leads to considerable thermal strains and stresses which need to be taken into account in the mechanical analysis. Using standard procedures, the variational equation for the mechanical problem in friction drilling process is written to find the displacement field $u(X, t) \in V = \{u \in H^1(\Omega) : u = u_g \text{ on } \partial\Omega_g\}$, such that for arbitrary variation $\delta u \in V_0 = \{u \in H^1(\Omega) : u = 0 \text{ on } \partial\Omega_g\}$, the following equation is satisfied:

$$\int_{\Omega} \rho \ddot{u} \cdot \delta u \, d\Omega + \int_{\Omega} \delta\Big(\varepsilon(u)\Big)^T : \sigma \, d\Omega = \int_{\Omega} b \cdot \delta u \, d\Omega + \int_{\Omega_h} h \cdot \delta u \, ds + \int_{\partial\Omega_c} \gamma \cdot \delta u \, ds \tag{3}$$

where b is the body force vector and σ is the Cauchy stress obtained from the constitutive law which is temperature dependent. The rate representation of strain field $\dot{\varepsilon}$ should consider the thermal effect which is described by

$$\dot{\varepsilon} = \dot{\varepsilon}^e + \dot{\varepsilon}^p + \dot{\varepsilon}^\theta \tag{4}$$

where $\dot{\varepsilon}^e$ is elastic strain rate tensor, and $\dot{\varepsilon}^\theta = \alpha\dot{\theta}$ is the thermal strain rate tensor with α denoting the thermal expansion coefficient. $\partial\Omega_g$ denotes a Dirichlet boundary imposed by a displacement u_g, and $\partial\Omega_h$ is the Neumann boundary prescribed by a surface traction h. γ denotes the contact traction which is governed by the unilateral contact conditions and Coulomb friction law [3]. Using Eq. (4) and the isothermal assumption from the staggered time-steeping algorithm, the corresponding rate form of the constitutive relation in mechanical phase can be written as

$$\dot{\sigma} = C(\theta) : \left(\dot{\varepsilon} - \dot{\varepsilon}^p - \dot{\varepsilon}^\theta\right) \tag{5}$$

where C is the temperature-dependent fourth-order isotropic elastic tensor.

Consequently, the thermal-mechanical problem in metal drilling process can be stated by coupling the mechanical weak form in Eq. (3) with the thermal weak form in Eq. (1) using the staggered time marching scheme. The coupled system of equations is discretized using meshfree approximations and solved by the classical explicit time-stepping approach which is described in the next section.

3 Particle Formulation

3.1 Meshfree Approximation and Discretization

The standard meshfree Galerkin method [8] for the thermal problem is formulated on a finite dimensional space $\Theta^s h \subset \Theta$ employing the thermal weak form of Eq. (1) to find $\theta^h(t) \in \Theta^h$ such that

$$\int_{\Omega} \rho C_p \dot{\theta}^h \delta\theta \, d\Omega + \int_{\Omega} k \nabla \theta^h \cdot \nabla\left(\delta\theta^h\right) d\Omega = \int_{\partial\Omega_n} q_n \delta\theta^h \, ds + \int_{\Omega} Q \delta\theta^h \, d\Omega$$
$$+ \int_{\partial\Omega_c} h_c \left(\theta^h - \theta_{\text{tool}}\right) \delta\theta^h \, ds + \int_{\partial\Omega_c} \eta\tau \cdot \left[\dot{u}^t\right] \delta\theta^h \, ds \quad \forall \delta\theta^h \in \Theta_0^h \tag{6}$$

with initial condition

$$\theta^h(X, 0) = \theta_0(X) \quad \text{in } \Omega \tag{7}$$

where and $\Theta^h = \text{span}\{\phi_I^a : I \in Z_I\}$ and Z_I is an index set. $\{\phi_I^a\}_I \in Z_I$ are meshfree shape functions constructed by the meshfree convex approximation [35, 36] which is employed in this study to simplify the boundary condition enforcement.

For a particle distribution denoted by an index set $Z_I = \{X_I\}_{I=1}^{NP} \subset \mathbb{R}^3$, approximating the displacement field using the meshfree approximation gives

$$u^h(X, t) = \sum_{I \in Z_I} \phi_I^a(X) u(X_I, t) = \sum_{I \in Z_I} \phi_I^a \tilde{u}(t) \quad \forall X \in \Omega \tag{8}$$

where NP is the total number of particles in discretization. $\phi_I^a(X)$, $I = 1, \ldots, NP$ can be interpreted as Lagrangian shape functions of the meshfree approximation for the displacement field u^h as well as the temperature field θ^h where the superscript "a" denotes the support size of $\phi_I^a(X)$.

In order to prevent the tensile instability caused by the Eulerian kernel, the Lagrangian kernel approach [8] is considered in this development. Correspondingly, Eq. (6) is rewritten by

$$\int_{\Omega} \rho C_p \dot{\theta}^h \delta\theta \, d\Omega + \int_{\Omega} \left(F^{-1} \cdot K \cdot F^{-T} \cdot \nabla^0 \theta^h\right) \cdot \nabla^0 \left(\delta\theta^h\right) d\Omega = \int_{\partial\Omega_n} q_n \delta\theta^h \, ds$$
$$+ \int_{\Omega} Q \delta\theta^h \, d\Omega + \int_{\partial\Omega_c} h_c \left(\theta^h - \theta_{\text{tool}}\right) \delta\theta^h \, ds + \int_{\partial\Omega_c} \eta\tau \cdot \left[\dot{u}^t\right] \delta\theta^h \, ds$$
$$\forall \delta\theta^h \in \Theta_0^h \tag{9}$$

where ∇^0 denotes the gradient operator with respect to reference position X, $K = kI^{(2)}$ is the thermal conductivity tensor with $I^{(2)}$ denoting the second-order identity tensor, and F is the deformation gradient. We remark that although the term *tensile instability* is reserved to describe the numerical instability of particle methods in structural analysis, we take the term in this paper to address the similar instability caused by the Eulerian kernel in the coupled thermo-mechanical analysis. Consequently, discrete points from meshfree discretization that carry the primary unknown variables are attached to the same set of material points throughout the course of deformation in Lagrangian particle methods. Under this consideration, the node set $Z_I = \{X_I, I = 1, \ldots, NP\}$ is the set of nodes defined in the reference configuration. In practice, the set of meshfree nodes can be taken from the finite element nodes created by a finite element mesh generator initially. Thus the geometrical representation of Ω can be numerically approximated by $\Omega \approx \bigcup_{I=1}^{NP} \Omega_I$ where Ω_I refers the volume of particle I which can be evaluated at time $t = 0$ using the information from the finite element mesh. The resultant discrete equations are then integrated using the direct nodal integration (DNI) scheme.

We can also formulate the mechanical weak form of Eq. (3) in similar fashion. Nevertheless, an application of the DNI scheme to Lagrangian particle methods leads to another numerical instability known as the zero-energy mode in structural analysis. To suppress the zero-energy mode and stabilize the solution for friction drilling simulation, the standard smoothed particle Galerkin (SPG) method [28, 29] is employed with a consideration of the thermal effect. The essence of SPG method in structural analysis is to augment the standard energy functional by a stabilization term using the penalty approach. As opposed to the residual-type stabilization method [22] which uses the residual of the momentum equation and artificial control parameters to effect stabilization, SPG method introduces a projection of displacement gradients on to a strain space leading to an additional term that penalizes the difference in strain fields for stabilization. The penalty approach modifies the DNI scheme and gives rise to a dual stress-points algorithm [28] which can be easily implemented and parallelized for the large-scale computation in industrial applications. The reader is refer to [28, 29] for detail information and references on SPG method. The SPG method for mechanical part of the coupling problem we considered is then as follows: find $u^h(X,t) \in V^h$ such that

$$\int_{\Omega} \rho \ddot{u}^h \cdot \delta u^h \, d\Omega + \underbrace{\int_{\Omega} \sigma : \left(F^{-1} \cdot \nabla^0 \delta u^h\right) d\Omega}_{\text{standard}} + \underbrace{\int_{\Omega} \delta\left(F^{-1} \cdot \nabla^0 \left(F^{-1} \cdot \nabla^0 \delta u^h\right) \cdot \lambda\right)^T : \tilde{\sigma} \, d\Omega}_{\text{stabilization}}$$
$$= \int_{\Omega} b \cdot \delta u^h \, d\Omega + \int_{\Omega_n} h \cdot \delta u^h \, ds + \int_{\partial\Omega_c} \gamma \cdot \delta u^h \, ds \quad \forall \delta u^H \in V_0^h \tag{10}$$

with initial conditions

$$u^h(X,0) = u_0(X) \tag{11}$$

$$\dot{u}^h(X,0) = \dot{u}_0(X) \tag{12}$$

where the stabilization term is composed of first-order strain gradients, stabilization stresses $\tilde{\sigma}$, and stabilization coefficient matrix $\lambda(x)$ that can be found in [28, 29].

3.2 Semi-discrete Equations

The semi-discrete equations of the thermal problem can be expressed by the following algebraic equations.

$$\tilde{\theta} + \mathbf{H}\tilde{\theta} = \mathbf{P} \tag{13}$$

where

$$C_{IJ} = \int_{\Omega_x} \rho_0 C_p \Psi_I \Psi_J \, d\Omega_X \tag{14}$$

$$H_{IJ} = \int_{\Omega} k F_{il}^{-1} F_{lj}^{-T} \Psi_{I,j} \Psi_{J,i} \, d\Omega + \int_{\partial\Omega_c} h_c \Psi_I \Psi_J \, ds \tag{15}$$

$$P_I = \int_{\Omega} \eta S : \dot{\varepsilon}^p \Psi_I \, d\Omega + \int_{\partial\Omega_n} q_n \Psi_I \, ds + \int_{\partial\Omega_c} \left(h_c \theta_{\text{tool}} - \boldsymbol{\eta} \tau \cdot \left[\dot{u}^t \right] \right) \Psi_I \, ds \tag{16}$$

Thermal equation in Eq. (13) is marched through time using the forward difference algorithm [3] which is given by

$$\tilde{\theta}_{n+1} = \tilde{\theta}_n + \Delta t \dot{\tilde{\theta}}_n \tag{17}$$

$$\dot{\tilde{\theta}}_n = {C^l}^{-1} \left(P_n - H_N \tilde{\theta}_n \right) \tag{18}$$

where the thermal capacity matrix C is advantageously replaced by the lumped matrix C^l for the explicit analysis. When the tool surface is in contact with the workpiece, the standard Fourier's law cannot be used to fully describe the heat transfer phenomena because the contact surfaces do not physically match perfectly. In this case, the heat resistance generally decreases as contact pressure increases. For this reason, the heat conductance h^c in the thermal contact is assumed to be a function of normal contact pressure, thermal conductivity of the gas, yielding stress of the work piece and surface roughness as described in [37].

In a similar way, the semi-discrete equations of the mechanical problem are given by

$$M\ddot{\tilde{U}} = F^{\text{ext}} + F^{c} - F^{\text{int}} - F^{\text{stab}} \tag{19}$$

where the mass matrix M, external force F^{ext}, internal force F^{int}, and stabilization force F^{stab} for the SPG method can be found in [28, 29, 32]. F^{c} is the contact force which is given by

$$F_I^c = \int_{\partial\Omega_c} \gamma \phi_I^a \, ds \tag{20}$$

Since the tungsten carbide tool is also meshed by the finite element discretization, the mechanical contact between the workpiece and drilling tool is modelled using the standard node-to-surface penalty contact algorithm [3, 38].

It also suffices to integrate Eq. (19) by the central difference integration algorithm and results in

$$\dot{\tilde{U}}_{n+1/2} = \dot{\tilde{U}}_{n-1/2} + \frac{\Delta t_{n+1} + \Delta t_n}{2} \ddot{\tilde{U}}_n \tag{21}$$

$$\dot{\tilde{U}}_{n+1} = \tilde{U}_n + \Delta t_{n+1} \ddot{\tilde{U}}_{n+1/2} \tag{22}$$

$$\ddot{\tilde{U}}_n = {M^l}^{-1} \left(F_n^{\text{ext}} + F_n^{c} - F_n^{\text{int}} - F_n^{\text{stab}} \right) \tag{23}$$

where M^l is the lumped mass matrix. Noting that the temperature remains constant and material properties are temperature dependent during this mechanical phase.

The critical time step in the explicit method is governed by the Courant-Friedrichs-Lewy (CFL) condition [3] which is given in the following for the thermal and mechanical analysis respectively

$$\Delta t^{\theta} \le S_c \min \left(\frac{\rho C_p l^2}{2k} \right), \quad \Delta t^{u} \le S_c \min \left(\frac{l}{C_u} \right) \tag{24}$$

where the sound speed C_u gives the characteristic speed of the medium in mechanical analysis. l is the support size of the kernel [8] for the particle system. A scaling factor $S_c = 0.15$ is used in this study.

4 Large Deformation and Material Separation Analyses

In this section, we discuss the computational procedures of the present method for the analysis of large deformation and material separation problems in frictional drilling simulation.

4.1 The Adaptive Anisotropic Lagrangian Kernel for Large Deformation Analysis

As mentioned earlier in the Introduction, Lagrangian kernel has been utilized in the particle method to remove the tension instability in the nonlinear structural analysis. However, the Lagrangian particle methods experience the excessive straining problem when the strictly use of Lagrangian kernel is no more applicable. Specifically, the excessive straining during the large deformation friction drilling simulation inevitably causes the numerical breakdown when the deformation gradient computed at the particle ceases to become invertible.

In order to handle the excessive straining problem, an adaptive anisotropic Lagrangian kernel is considered [28]. Using the chain rule, the calculation for the deformation gradient at the particle can be rewritten [32] as

$$F^{n+m} = \widehat{F}^{n+m} F^{n} \tag{25}$$

where $\widehat{F}^{n+m}(\widehat{x})$ is the decomposed deformation gradient, from $t = t_n$ to t_{n+m}, computed based on the new reference configuration and is given by

$$\begin{aligned} \widehat{F}_{ij}^{n+m}(X_J) = & \frac{\partial \widehat{x}_i}{\partial \widehat{X}_j} = \sum_{I=1}^{NP} \frac{\partial \phi_I^a(\widehat{X}_J)}{\partial \widehat{X}_j} \widehat{x}_{iI}(X, t_{n+m}) \\ = & \sum_{I=1}^{NP} \frac{\partial \phi_I^a(\widehat{X}_J)}{\partial \widehat{X}_j} \left(\widehat{X}_{iI} + \tilde{u}_{iI}(X, t_{n+m}) \right) \\ = & \delta_{ij} + \sum_{I=1}^{NP} \frac{\partial \phi_I^a(\widehat{X}_J)}{\partial \widehat{X}_j} \tilde{u}_{iI}(X, t_{n+m}) \end{aligned} \tag{26}$$

Here, $\widehat{x} = \widehat{X} + \tilde{u}(X, t_{n+m})$ is a position vector defined in the new reference configuration $\widehat{X} = x(X, t_n)$. A local $\widehat{x}^I$-coordinate system in which the axes are parallel to the global Cartesian coordinates and the origin is located at $\widehat{x}_I$ which is defined for each particle I. In each new reference configuration, an ellipsoidal nodal support is defined for the neighbor particle searching. The three-dimensional ellipsoidal cubic spline kernel function is defined in another local $\widehat{\widehat{x}}^I$-coordinate system by

$$\varphi_I^a(\widehat{X}_J) = \varphi_1\left(\frac{\widehat{\widehat{X}}_J^I}{h_1^n}\right)\varphi_1\left(\frac{\widehat{\widehat{Y}}_J^I}{h_2^n}\right)\varphi_1\left(\frac{\widehat{\widehat{Z}}_J^I}{h_3^n}\right) \tag{27}$$

where ϕ_1 is a standard one-dimensional cubic spline kernel function, h_1^n, h_2^n and h_3^n are the current semi-major axes of the ellipsoid. The sizes of semi-major axes can be considered the support sizes of the kernel and are updated according to the deformation [28]. $\widehat{\widehat{x}}_J^I$, $\widehat{\widehat{y}}_J^I$ and $\widehat{\widehat{z}}_J^I$ are the projections of relative position vector $\widehat{x}_J - \widehat{x}_I$ on the local $\widehat{\widehat{x}}^I$-coordinate system respectively. The adaptive anisotropic Lagrangian kernel is updated constantly over a period of time. The spherical shape domain of cubic spline kernel function deforms and rotates according to the Lagrangian motion between each two adaptive Lagrangian kernel steps. We address the reader to reference [28] for a comprehensive description of the approach. For the computational efficiency in explicit time integration method, the material derivatives of meshfree shape functions are always computed and stored at the new reference configuration and reused during the time stepping.

Since the operation of adaptive anisotropic Lagrangian kernel does not involve remeshing, the stress-recovery techniques or remapping procedures are not necessary. This unique property of present method leads to a relatively simple mathematical formulation for simulating the large strain problems.

4.2 The Bond-Based Failure Criterion for Material Separation Analysis

Excessive straining also appears in the friction drilling process when the material of workpiece starts to fail at the bushing forming stage. Precisely, the C^1-continuity assumption in Lagrangian particle methods is inadequate to describe the kinematic discontinuity of displacement field in a continuous setting for the failure analysis [19]. This makes Lagrangian particle methods even more challenging in friction drilling simulation.

To further avoid the excessive straining problem due to the assumption of continuous displacement field in the friction drilling simulation, a bond-based failure criterion [29, 31] is incorporated with the present coupled thermal-mechanical formulations. The origins of this approach can be traced back to the bond failure in peridynamics [31, 39] in which material failure is modeled through bond breakage. In Lagrangian particle methods, the bond is a representation of a connection between two particles. Given a length of the bond $\|X_J - X_I\|$ for a particle pair consisting of particles I and J in the initial configuration, the stretch ratio e_{IJ} of the bond is defined by

$$e_{IJ} = \frac{\|x_J - x_I\|}{\|X_J - X_I\|} \tag{28}$$

For the friction drilling simulation, we restrict our attention to the material failure in metals. In the bond-based failure criterion for ductile material, two neighbor particles are considered disconnected during the neighbor particle sorting whenever their averaged effective plastic strain and stretch ratio reach their respective critical values. Accordingly, the three-dimensional ellipsoidal cubic spline kernel function in Eq. (27) for a pair of particles I and J can be modified as:

$$\phi_I^a\left(\widehat{X}_J\right) = \begin{cases} 0, & \text{if } \widehat{X}_J \notin \text{supp}\left(\phi_I^a\right) \\ & \text{or } \left(\bar{\varepsilon}_{IJ}^P > \bar{\varepsilon}_{\text{crit}}^p \text{ and } e_{IJ} > e_{\text{crit}}\right) \\ \phi_1\left(\frac{\widehat{\overline{X}}_J^I}{h_1^n}\right)\phi_1\left(\frac{\widehat{\overline{Y}}_J^I}{h_2^n}\right)\phi_1\left(\frac{\widehat{\overline{Z}}_J^I}{h_3^n}\right), & \text{otherwise} \end{cases} \tag{29}$$

where $\bar{\varepsilon}_{ij}^p = \left(\bar{\varepsilon}^p\left(\widehat{X}_J\right)\right)/2$, and $\bar{\varepsilon}^p$ denoting the effective plastic strain. $\bar{\varepsilon}_{\text{crit}}^p$is the critical effective plastic strain for bond failure, and e_{crit} denotes the critical stretch ratio. We consider $e_{\text{crit}} \geq 1.0$ in our numerical analysis which implies that the bond failure does not occur under compression. This implication is valid for most metal failure process.

Because the effective plastic strain at each particle is monotonically increasing during the course of deformation, the kinematic disconnection in a particle pair is considered as a permanent and irreversible process. This is a substantial characteristic for the present method in metal failure analyses since the non-physical material self-healing and excessive straining issues can also be completely excluded from the material failure simulation.

5 Numerical Example

A friction drilling process of AISI 304 stainless steel is modeled and compared with the experimental data in this section. The normalized nodal support size of 1.9 is used and the adaptive anisotropic Lagrangian kernel is updated every 50 time steps in the explicit dynamic analysis.

The AISI 304 stainless steel specimen used in the friction drilling process has a diameter of 18 mm and thickness of 1.5 mm [40]. The geometry of the tool is shown in Fig. 1a. The tool, which rotates at 3000 rpm and plunges at 100 mm/min in the test, is modeled by rigid material and meshed using tetrahedral finite elements. As can be seen in Fig. 1b, the metal workpiece is discretized using 12,607 Lagrangian particles. Finer discretization with a nodal distance of approximately 0.25 mm is employed in the central portion of the specimen where large deformation and material separation occurs. As such, the explicit time step size for thermal analysis is 50 μs and for structural analysis is about 4 μs. The perimeter of the workpiece is clamped. The stress flow in the AISI 304 steel is modeled by the Johnson-Cook material law [41] (parameters: $A = 205$ MPa, $B = 802.5$ MPa, $C = 0.08$, $m =$

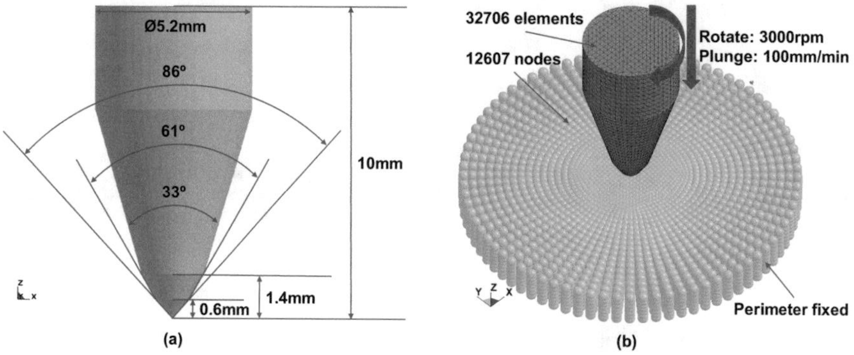

Fig. 1 Friction drilling: (**a**) tool geometry, (**b**) discretization and boundary conditions

1.09, $n = 0.622$). The failure behavior of the steel is handled by the SPG bond failure mechanism as described in Sect. 4.2 rather than the Johnson-Cook damage law, and the effective plastic strain for bond failure is set to 0.4. According to efunda (www.efunda.com), the Young's modulus of the workpiece is set to 193 GPa. The thermal properties of the AISI stainless steel are: coefficient of thermal expansion 0.0000184, heat capacity C_p 500 J/kg-K, and thermal conductivity k 16.2 W/m-K. The coefficient of friction (COF) between the tool and the workpiece is set to 0.2 for the node-to-surface contact algorithm in the numerical analysis. The fraction of heat generation η in the frictional contact is taken to be 0.5. The interfacial heat transfer between the tool and the workpiece is neglected. The Taylor-Quinney [34] coefficient η of 0.9 is considered in Eq. (2).

The comparison of thrust force and torque is presented in Fig. 2a and b, respectively. Both the force and torque responses capture the basic profiles of experimental data nicely. Further improvement in the force and torque results can be made by tuning the coefficient of contact friction. But this is not within the scope of this study and therefore not considered in this numerical example.

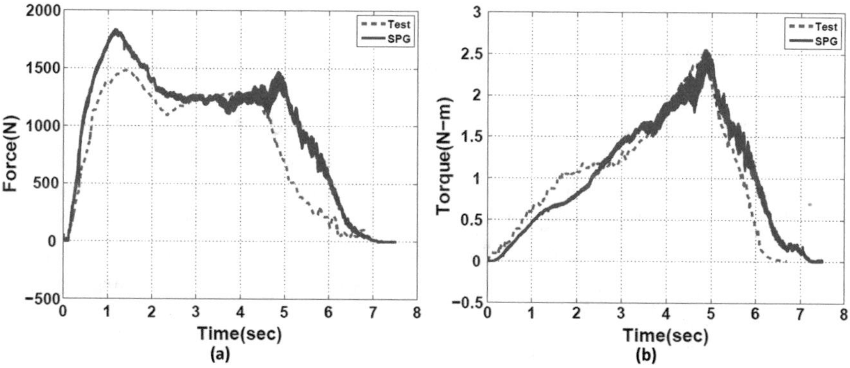

Fig. 2 Response of friction drilling: (**a**) thrust force, (**b**) torque

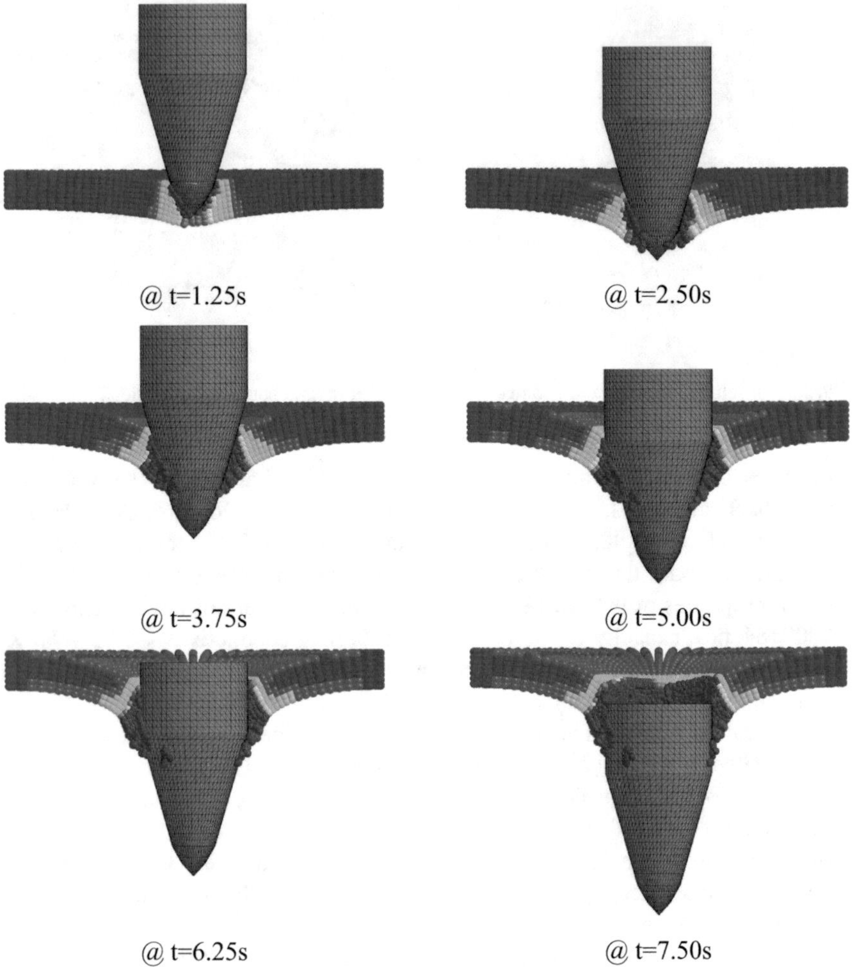

Fig. 3 Friction drilling: effective plastic strain distribution (red $\geq$ 0.4, blue: 0)

Figure 3 shows the evolution of the effective plastic strain in the workpiece while only half of it is plotted. Red color indicates effective plastic strain level of 0.4 (which is the bond failure criterion) or more. It should be pointed out that bond failure, i.e., material separation, only occurs when the effective plastic strain and stretch ratio both reach their respective critical values. As shown in Fig. 3, material failure occurs in a relatively small region and the bushing is qualitatively formed. It is worthwhile to emphasize that the creation of bushing shape is one of the major purposes of this type manufacturing process. However, it is not captured by any other numerical technique by far.

Figure 4 shows the simulation result of temperature distribution during the friction drilling process (back view). Red color indicates temperature rising of

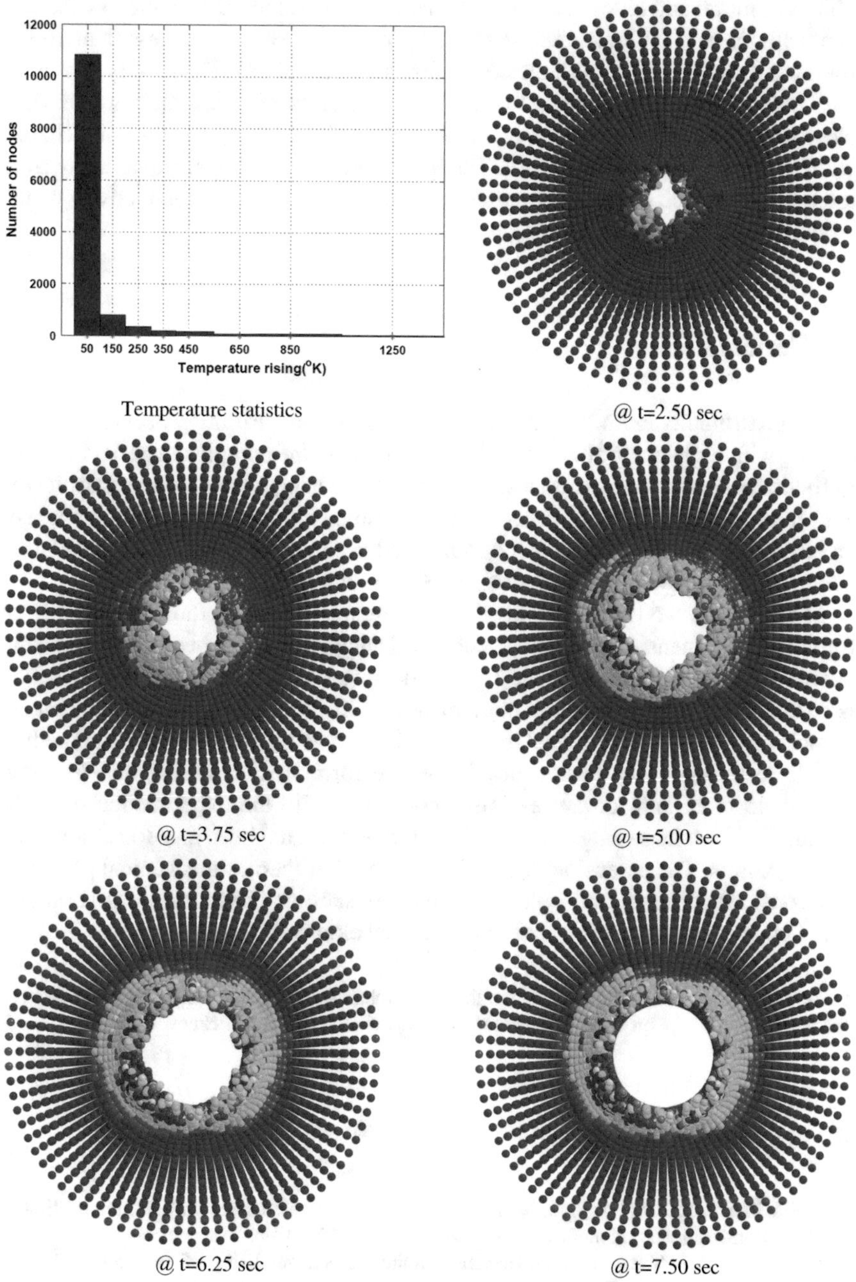

Fig. 4 Friction drilling: temperature field in workpiece (red: ≥ 100° K, blue: 0° K)

100°K or more, and blue color means no temperature rising. Due to the low thermal conductivity in AISI 304 stainless steel, the temperature rising is less than 100°K for more than 90% of the particles. Very few particles (about 1%) directly in contact with the tool have a temperature rising to 800°K or more. The simulation result is close to the measured temperature on the upper side of the disc at the contact zone which was reported at 842°K [40]. The heat wave did not propagate far away from the tool-workpiece contact region because of the low thermal conductivity of the workpiece and the fast machining process.

6 Conclusions

The main difficulty in finite element modeling of friction drilling process consists in dealing with high levels of deformations involving in the complex material flow due to frictional heating and material separation at the busing forming stage. Despite the enormous progress achieved lately in computational mechanics, the development of an advanced numerical tool for the robust and accurate friction drilling simulation continues to be nowadays an emerging need for industry.

In this study, we have introduced a Lagrangian particle method that is suitable for the three-dimensional thermo-mechanical analysis and can become a promising alternative numerical tool for the friction drilling simulation. The present method is developed to improve several numerical instabilities in conventional Lagrangian particle methods. The numerical results in this study suggest that the present method is able to produce the desired physics in the forming of a busing and generate reasonable force and torque responses compared with the experimental data. To the authors' best knowledge, the existing literature has not been able to demonstrate similar results. The extension of this method to other thermo-mechanical problems that consider complex multi-physics behaviors such as phase transformation and phase change will be the focus of our future development.

Acknowledgements The authors wish to thank Dr. John O. Hallquist of LSTC for his support to this research. The support from Ford Motor Company is also gratefully acknowledged.

References

1. S.F. Miller, A.J. Shih, P.J. Blau, Microstructural alterations associated with friction drilling of steel, aluminum, and titanium. J. Mater. Eng. Perform. **15**, 647–653 (2005)
2. S.F. Miller, A.J. Shih, Thermo-mechanical finite element modeling of the friction drilling process. J. Manuf. Sci. Eng. **129**, 531–538 (2007)
3. T. Belytschko, W.K. Liu, B. Moran, K.I. Elkhodary, *Nonlinear Finite Elements for Continua and Structures*, 2nd edn. (Wiley, Chichester, 2014)
4. B. Padma Raju, M. Kumara Swamy, Finite element simulation of a friction drilling procedure using Deform-3D. Int. J. Eng. Res. Tech. **2**, 716–721 (2012)

5. A. Gopichand, M. Veera Brahmam, D. Bhanuprakash, Numerical simulation and analysis of friction drilling process for aluminum alloy using Ansys. Int. J. Eng. Res. Tech. **3**, 602–607 (2014)
6. G. Buffa, J. Hua, R. Shivpuri, L. Fratini, A continuum based FEM model for friction stir welding – model development. Mater. Sci. Eng. A **419**, 389–396 (2006)
7. B. Meyghani, M.B. Awang, S.S. Emamian, M.K.B. Mohd Nor, S.R. Pedapati, A comparison of different finite element methods in the thermal analysis of friction stir welding (FSW). Metals **10**, 450 (2017)
8. J.S. Chen, C. Pan, C.T. Wu, W.K. Liu, Reproducing kernel particle methods for large deformation analysis of non-linear structures. Comput. Methods Appl. Mech. Eng. **139**, 195–227 (1996)
9. T. Rabczuk, T. Belytschko, Cracking particles: a simplified meshfree method for arbitrary evolving cracks. Int. J. Numer. Methods Eng. **61**, 2316–2343 (2004)
10. D.C. Simkins, S. Li, Meshfree simulation of thermo-mechanical ductile fracture. Comput. Mech. **38**, 235–249 (2006)
11. S. Li, W. K. Liu, *Meshfree Particle Method* (Springer, Berlin, 2004)
12. D.D. Wang, J.S. Chen, Locking free stabilized conforming nodal integration for meshfree Mindlin-Reissner plate formulation. Comput. Methods Appl. Mech. Eng. **193**, 1065–1083 (2004)
13. R.A. Gingold, J.J. Monaghan, Smoothed particle hydrodynamics – theory and application to non-spherical stars. Mon. Not. R. Astron. Soc. **181**, 375–389 (1977)
14. L.B. Lucy, A numerical approach to the testing of the fission hypothesis. Astron. J. **82**, 1013–1024 (1977)
15. L.D. Libersky, A.G. Petschek, Smooth particle hydrodynamics with strength of materials. Lect. Notes Phys. **395**, 248–257 (1990)
16. M.S. Shadloo, G. Oger, D.L. Touzé, Smoothed particle hydrodynamics method for fluid flows, towards industrial applications: motivations, current state, and challenges. Comput. Fluid **136**, 11–34 (2016)
17. W. Swegle, D.L. Hicks, S.W. Attaway, Smoothed particle hydrodynamics stability analysis, Comput. Mech. **116**, 123–134 (1995)
18. T. Belytschko, Y. Guo, W.K. Liu, S.P. Xiao, A unified stability analysis of meshless particle methods. Int. J. Numer. Methods Eng. **48**, 1359–1400 (2000)
19. C.T. Wu, N. Ma, K. Takada, H. Okada, A meshfree continuous-discontinuous approach for the ductile fracture modeling in explicit dynamics analysis. Comput. Mech. **58**, 391–409 (2016)
20. T. Rabczuk, T. Belytschko, S.P. Xiao, Stable particle methods based on Lagrangian kernels. Comput. Methods Appl. Mech. Eng. **193**, 1035–1063 (2004)
21. C.T. Dyka, P.W. Randles, R.P. Ingel, Stress points for tension instability in SPH. Int. J. Numer. Methods Eng. **40**, 2325–2341 (1997)
22. S. Beissel, T. Belytschko, Nodal integration of the element-free Galerkin method. Comput. Methods Appl. Mech. Eng. **139**, 49–74 (1996)
23. C.T. Wu, M. Koishi, W. Hu, A displacement smoothing induced strain gradient stabilization for the meshfree Galerkin nodal integration method. Comput. Mech. **56**, 19–37 (2015)
24. M. Hillman, J.S Chen, An accelerated, convergent, and stable nodal integration in Galerkin meshfree methods for linear and nonlinear mechanics. Int. J. Numer. Methods Eng. **107**, 603–630 (2016)
25. J.S. Chen, C.T. Wu, S. Yoon, Y. You, A stabilized conforming nodal integration for Galerkin Meshfree methods. Int. J. Numer. Methods Eng. **50**, 435–466 (2001)
26. M. Hillman, J.S. Chen, S.W. Chi, Stabilized and variationally consistent nodal integration for meshfree modeling of impact problems. Comput. Part. Mech. **1** 245–256 (2014)
27. P.C. Guan, J.S. Chen, Y. Wu, H. Tang, J. Gaidos, K. Hofstetter, M. Alsaleh, Semi-Lagrangian reproducing kernel formulation and application to modeling earth moving operations. Mech. Mater. **41**, 670–683 (2009)

28. C.T. Wu, S.W. Chi, M. Koishi, Y. Wu, Strain gradient stabilization with dual stress points for the meshfree nodal integration method in inelastic analysis. Int. J. Numer. Methods Eng. **107**, 3–30 (2016)
29. C.T. Wu, Y. Wu, J.E. Crawford, J.M. Magallanes, Three-dimensional concrete impact and penetration simulations using the smoothed particle Galerkin method. Int. J. Impact Eng. **106**, 1–17 (2017)
30. J.S. Chen, X. Zhang, T. Belytschko, An implicit gradient model by a reproducing kernel strain regularization in strain localization problems. Comput. Methods Appl. Mech. Eng. **193**, 2827–2844 (2014)
31. S.A. Silling, E. Askari, A meshfree method based on the peridynamic model of solid mechanics. Comput. Struct. **83**, 1526–1535 (2005)
32. C.T. Wu, T.Q. Bui, Y.C. Wu, T. L. Luo, M. Wang, C.C. Liao, P.Y. Chen, Y.S. Lai, Numerical and experimental validation of a particle Gakerkin method for metal grinding simulation. Comput. Mech. **61**(3), 365–383 (2018). https://doi.org/10.1007/s00466-017-1456-6
33. C.A. Felippa, K.C. Park, Staggered transient analysis procedures for coupled dynamic systems. Comput. Methods Appl. Mech. Eng. **26**, 61–112 (1980)
34. T. J. R. Hughes, *The Finite Element Method* (Prentice-Hall, Englewood Cliffs, 2000)
35. C.T. Wu, C.K. Park, J.S. Chen, A generalized approximation for the meshfree analysis of solids. Int. J. Numer. Methods Eng. **85**, 693–722 (2011)
36. C.T. Wu, M. Koishi, Three-dimensional meshfree-enriched finite element formulation for micromechanical hyperelastic modeling of particulate rubber composites. Int. J. Numer. Methods Eng. **91**, 1137–1157 (2012)
37. I.T. Shvets, E.P. Dyban, Contact heat transfer between plane metal surfaces. Int. Chem. Eng. **12**, 621–624 (1964)
38. J.O. Hallquist, *LS-DYNA Theory Manual* (Livermore Software Technology Corporation, Troy, 2006)
39. B. Ren, C.T. Wu, E. Askari, A 3D discontinuous Galerkin finite element method with the bond-based peridynamics model for dynamics brittle failure analysis. Int. J. Impact Eng. **99**, 14–25 (2017)
40. P. Krasauskas, S. Kilikevičius, R. Česnavičius, D. Pačenga. Experimental analysis and numerical simulation of the stainless AISI 304 steel friction drilling process. Mechanika **20**, 590–595 (2014)
41. G.R. Johnson, W.H. Cook, A constitutive model and data for metals subjected to large strains, high strain rates, and high temperatures, in *Proceedings of the 7th International Symposium on Ballistics*, The Hague, 19–21 April 1983, pp. 541–547

Global-Local Enrichments in PUMA

Matthias Birner and Marc Alexander Schweitzer

Abstract In this paper we present the global-local enrichment approach in a general partition of unity method. Moreover, we propose an automatic scheme of computing an optimal parameter in Robin boundary conditions for the local problem. We present results of two dimensional fracture mechanics problems to demonstrate the properties and performance of the resulting method.

1 Introduction

Numerical simulation of fracture mechanics problems is a widely used tool in industry to predict the damage tolerance of a mechanical structure. To this end, the finite element method (FEM) [3] has been used for decades in industrial scale problems [20, 32]. Yet multi-scale problems, like small cracks in a relatively big geometry, are to the disadvantage of FEMs: As the number of cracks increases, the required computational power explodes [14]. This is due to the need of a strongly refined mesh around the crack-fronts, in order to capture the singularities arising there. Furthermore, the mesh has to be regenerated on every load-step when solving crack growth problems [31], further increasing the time-to-solution. Partition of unity methods [5, 16, 21] can overcome both drawbacks by being able to incorporate

M. Birner (✉)
Fraunhofer-Institut für Algorithmen und Wissenschaftliches Rechnen SCAI, Sankt Augustin, Germany
e-mail: matthias.birner@scai.fraunhofer.de

M. A. Schweitzer
Fraunhofer-Institut für Algorithmen und Wissenschaftliches Rechnen SCAI, Sankt Augustin, Germany

Rheinische Friedrich-Wilhelms-Universität Bonn, Institut für Numerische Simulation, Bonn, Germany
e-mail: marc.alexander.schweitzer@scai.fraunhofer.de; schweitzer@ins.uni-bonn.de

M. Griebel, M. A. Schweitzer (eds.), *Meshfree Methods for Partial Differential Equations IX*, Lecture Notes in Computational Science and Engineering 129,
https://doi.org/10.1007/978-3-030-15119-5_10

arbitrary basis functions, thereby requiring only a constant number of degrees of freedom to resolve any crack or other localized feature, independent of its size and geometry. This is achieved by having a partition of unity that covers the computational domain and multiplying local approximation spaces to each partition of unity function. These local spaces usually consist of a smooth, polynomial part—as in finite elements—and an enrichment part, that includes problem specific functions. In the simulation of fracture mechanics problems, we want to include enrichments that resolve the singularities and discontinuities arising in vicinity of the cracks. Unfortunately, analytic knowledge of such basis functions is limited to two dimensions. A recent method to circumvent this limitation are global-local enrichments [4, 7, 13, 15, 19], where the fundamental idea is to compute problem specific basis functions on the fly during the simulation. This is achieved by setting up local problems around geometric features with boundary data from an initial, coarse global solution, that disregards those features. The solutions of these local problems are then used as enrichment functions on the global problem, allowing us to keep the global discretization coarse, while retaining resolution of fine scale details. By this, we arrive at an approximation that has high local accuracy with only adding a few degrees of freedom, as opposed to applying (adaptive) h-refinement towards the crack [14].

There already exists an implementation of, and research on global-local enrichments in the finite element based partition of unity method (GFEM^{gl}) [7] by Duarte and Babuška. In this paper we transfer those ideas to the meshfree and more general partition of unity method (PUM) [21] and study the construction of optimal boundary conditions for the local problems in this context, see also [4, 13]. The fundamental assumption here is, that even though the initial global solution does not capture the fine scale behavior near the cracks, it should be accurate enough further away from them. Hence, applying a buffer zone [4], i.e., computing the local enrichment functions on a larger subdomain compared to the area the enrichment is used on, is a first approach to moderate the influence of inexact boundary data. Another approach to improve the boundary data is to use multiple global-local iterations [13], as this drastically improves the quality of the boundary data. Furthermore, different kinds of boundary conditions imposed on the local problems may vary in their sensitivity to inexact boundary data and material properties.

In the following we discuss these aspects of global-local enrichments in the PUM context. The remainder of this paper is structured as follows. First, we shortly review the PUM in Sect. 2.1 and the global-local enrichment scheme in Sect. 2.2. The considered model problem and its respective formulation for the global-local enrichment approach is presented in Sect. 3. Here we discuss the various choices to improve the boundary data for the local-problem and propose a simple yet very effective constructive approach to the identification of optimal parameters for Robin type boundary conditions. In Sect. 4 we present the results of our numerical experiments which clearly show the desired behavior. We conclude with some remarks in Sect. 5.

2 Prerequisites

2.1 Partition of Unity Method

Partition of unity methods (PUM) are a class of methods to numerically solve partial differential equations (PDEs), introduced to overcome limitations in the choice of basis functions [16] of classical finite element methods (FEM). The key concept of PUMs is the use of a compactly supported partition of unity (PU), that covers the computational domain Ω. To each PU function φ_i a local approximation space V_i is attached, which yields a global, finite-dimensional space, that is then used in a Galerkin approach. An advantage over classical FEMs is the ability to incorporate arbitrary basis functions, where the intent is to use only a few problem specific basis functions, thereby requiring less degrees of freedom to attain the desired global accuracy. Well-known instances of the PUM are the generalized finite element method (GFEM) by Duarte and Babuška [5] and the extended finite element method (XFEM) by Moës, Dolbow and Belytschko [17]. The particular PUM employed in this study was introduced in [11, 12, 21]. All computations in this study were carried out using the PUMA software framework [9, 27] developed at Fraunhofer SCAI. In the following we present only a very short review of this PUM and refer the reader to [21] for details.

Given a computational domain Ω, we assume to have a partition of unity $\{\varphi_i\}$ with $\varphi_i \geq 0$ and

$$\sum_i \varphi_i(x) = 1 \quad \forall x \in \Omega \tag{1}$$

that covers the domain. We call the support of a partition of unity function φ_i a patch $\omega_i := \operatorname{supp}(\varphi_i)$. In the PUMA software framework, patches are constructed as follows. First, we compute a cubic bounding-box C of the domain Ω. This bounding box corresponds to discretization level zero. For level l, we sub-divide the bounding-box l-times (uniformly) and obtain the cells

$$C_i = \prod_{k=1}^{d} \left(c_i^k - h, c_i^k + h\right), \tag{2}$$

where the c_i are the mid-points of the cells. We then obtain the patches ω_i by scaling the cells via

$$\omega_i := \prod_{k=1}^{d} \left(c_i^k - \alpha h, c_i^k + \alpha h\right), \qquad \alpha > 0. \tag{3}$$

On these patches ω_i we construct a Shepard partition of unity with the help of B-spline weight functions. Details on this construction can be found in e.g. [21, 22].

To construct a higher order basis, each PU function φ_i is multiplied with a local approximation space

$$\mathrm{V}_i = \mathcal{P}_i \oplus \mathcal{E}_i = \mathrm{span}\langle \psi_i^s, \eta_i^t \rangle, \tag{4}$$

of dimension n_i, where $\mathcal{P}_i$ are spaces of polynomials of degree $d^{\mathcal{P}_i}$, and the $\mathcal{E}_i$ denote so-called enrichment spaces of dimension $d^{\mathcal{E}_i}$. The latter are arbitrary functions locally incorporated into the simulation, which we can choose with respect to the problem at hand. We either obtain the $\mathcal{E}_i$ from a-priori analytic knowledge about the structure of the solution to a problem, as in $2D$ fracture mechanics problems, where we have an analytic expansion of the solution around a crack tip available. Or the enrichments themselves are results of other simulations, as e.g. proposed in [1, 2, 26, 28] or in the global-local enrichment method [7] we discuss here. The global approximation space then reads

$$\mathrm{V}^{PU} = \sum_i \varphi_i \mathrm{V}_i = \sum_i \varphi_i \mathcal{P}_i + \varphi_i \mathcal{E}_i. \tag{5}$$

Observe, that we do not assume the discrete functions

$$\mathrm{u}^{PU} = \sum_i \varphi_i \left(\sum_{s=1}^{d^{\mathcal{P}_i}} \mathrm{u}_i^s \, \psi_i^s + \sum_{t=1}^{d^{\mathcal{E}_i}} \mathrm{u}_i^{t+d^{\mathcal{P}_i}} \, \eta_i^t, \right) \tag{6}$$

i.e. the employed basis functions, to be interpolatory. That is, the coefficients u_i^j do not necessarily correspond to function values at specific points. We therefore use the direct splitting of the local spaces V_i presented in [22] to enforce Dirichlet boundary conditions.

2.2 Numerical Enrichment Functions

The PUM attains its approximation properties from the local approximation spaces V_i. Thus, the choice of good approximation spaces is essential. If there is a-priori analytic knowledge about the solution available, the choice of V_i is rather simple, compare [25] and the references therein. However, the design of good local approximation spaces without analytic information, i.e. by numerical computation is an active research field and a number of different approaches have been proposed, see e.g. [26] and the references therein. The global-local enrichment technique [7] is such an approach, which we summarize in the following.

Preceding global-local enrichments in the GFEM was the attempt to compute more universally applicable enrichments in [28]. There, Strouboulis et al. computed enrichments by solving local problems for different predefined boundary conditions that should mimic general load cases. Yet, the quality of the computed enrichment

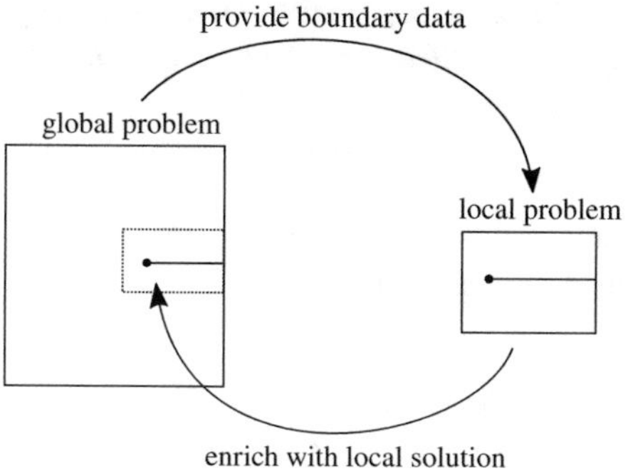

Fig. 1 The basic global-local enrichment cycle. The global problem we ultimately want to solve and a local problem set up around a crack. We first solve the (initial) global problem disregarding the crack, to provide boundary data for the local problem. The solution of the local problem is then used as a basis function on the global problem. Finally we solve the global problem again

then depends heavily on the prescribed boundary conditions, where there is no analytical answer on which to choose. In [7] the approach was to compute more problem related enrichments using boundary data from the global problem, thereby introducing the GFEM^{gl}.

An ancestor of global-local enrichments is the global-local finite element analysis (FEM^{gl}) [18] introduced in the 1970s, by which we specifically refer to the zooming technique [8, 30]. The FEM^{gl} addresses two-scale problems, where we have a small area of interest, e.g. a crack, in a relatively large domain. To account for this, the FEM^{gl} separates the scales, by first solving the global problem on a rather coarse mesh, and then setting up a finer local problem in the area of interest, where the global solution is used as boundary data. The final solution then is obtained from the direct combination of the global and the local solution.

Note that this procedure does not allow for any feedback from the local solution to the global problem. An approach that accounts for such feedback are global-local enrichments (GFEM^{gl}) as introduced in [7] for the generalized finite element method (GFEM) by Duarte and Babuška. The GFEM^{gl} achieves this by incorporating the local solution—computed just as in the FEM^{gl}—as an enrichment function into the global basis and solving the now enriched global problem again, compare Fig. 1. We shortly review the precise problem formulation in the following Section.

3 Model Problem and Global-Local Formulation

To this end, let us first introduce our general model problem, the equations of linear elasticity. On the global domain Ω_G we consider

$$\nabla \cdot \sigma = f \qquad \text{in } \Omega_G, \tag{7}$$

where f are the volume forces acting on the body, e.g. gravity, and σ is the Cauchy stress tensor. The latter is computed from the linear strain tensor ε via Hooke's law

$$\sigma = C : \varepsilon, \tag{8}$$

where Hooke's tensor C represents material parameters and the strain tensor is computed from the displacement field u by

$$\varepsilon(\mathrm{u}) = \frac{1}{2}\left(\nabla \mathrm{u} + (\nabla \mathrm{u})^T\right). \tag{9}$$

To obtain a unique solution of (7) we impose boundary conditions

$$\begin{aligned} \mathrm{u} &= \bar{\mathrm{u}} && \text{on } \Gamma_G^D \subset \partial\Omega_G \\ \sigma \cdot n &= \bar{t} && \text{on } \Gamma_G^N = \partial\Omega_G \backslash \Gamma_G^D, \end{aligned} \tag{10}$$

where n is the outward unit normal to Γ_G^N and $\bar{\mathrm{u}}$ and $\bar{t}$ are the prescribed displacement and traction, respectively.

The weak formulation of the global problem is then given by: Find $\mathrm{u}_G \in \mathrm{V}_G^{\mathrm{PU}} \subset H^1_{\Gamma_G^D}(\Omega_G)$ such that

$$\int_{\Omega_G} \sigma(\mathrm{u}_G) : \varepsilon(\mathrm{v}_G)\, \mathrm{d}x = \int_{\Gamma_G^N} \bar{t}\mathrm{v}_G\, \mathrm{d}s + \int_{\Omega_G} f\, \mathrm{v}_G\, \mathrm{d}x \tag{11}$$

for all test functions $\mathrm{v}_G \in \mathrm{V}_G^{\mathrm{PU}}$ that vanish on Γ_G^D, see [22] for details.

The global domain may have several areas in which we expect the solution to have particularly fine grained behavior, like cracks, corners or holes. To improve the approximation, we now set up a local problem for each such feature on a respective subdomain $\Omega_L \subset \Omega_G$, compare Fig. 1. On this local domain, we now define a local problem by restricting the global problem (7), i.e. (11), to the local domain Ω_L. Wherever possible we use the provided global boundary data (10) to define boundary conditions for the local problem.

The boundary of the local domain $\Gamma_L := \partial\Omega_L$ however, can consist of two parts: The, possibly empty, intersection with the global boundary $\Gamma_G := \partial\Omega_G$ and its complement $\Gamma_L \backslash \Gamma_G$ in the interior of the global domain Ω_G. On the former we apply the given boundary conditions of the global problem (10). And on the latter we use the computed global solution u_G to define boundary data for the local problem. Here we can choose whether to prescribe Dirichlet, Neumann or Robin boundary conditions, where the latter depend on a parameter $\kappa \geq 0$ and "interpolates" between the first two types. Thus, we focus on Robin boundary conditions in the following. For our local problem they read as

$$t(\mathrm{u}_L) + \kappa \mathrm{u}_L = t(\mathrm{u}_G) + \kappa \mathrm{u}_G \qquad \text{on } \Gamma_L \backslash \Gamma_G, \tag{12}$$

where $t(\mathrm{u}) := \sigma(\mathrm{u}) \cdot n$ denotes the traction associated with a displacement field u. Observe that for $\kappa = 0$ we get Neumann boundary conditions, and for large κ we impose Dirichlet boundary conditions by (12).

Research on the GFEM^{gl} in [14] finds that Robin boundary conditions in general yield the best overall approximation. Furthermore, work on non-overlapping Schwarz methods [10] suggests that Robin boundary conditions yield optimal approximation results, especially with no or only small buffer zones. Yet, all this depends on finding the right parameter κ. To this end, we propose the following simple scheme.

Note that one term of the right-hand side of (12) contains the traction $t(\mathrm{u}_G)$, i.e. the gradient $\nabla \mathrm{u}_G$ and the material parameters, where the other term depends on the global solution u_G and κ. From a stability point of view, we want both terms, or ultimately both assembled vectors in the linear system, to be of comparable size. Hence, we need to balance the global solution's values with its traction, which depends on the gradient of the global solution and further scales with the material parameters. To this end, we assemble the vectors

$$M_T(\mathrm{u}_G) = \left(\int\limits_{\Gamma_L \setminus \Gamma_G} t(\mathrm{u}_G)\mathrm{v}_i \,\mathrm{d}s \right)_{i=1}^{n} \quad \text{and} \quad M_D(\mathrm{u}_G) = \left(\int\limits_{\Gamma_L \setminus \Gamma_G} \mathrm{u}_G \mathrm{v}_i \,\mathrm{d}s \right)_{i=1}^{n} \tag{13}$$

where $\{\mathrm{v}_i\}_{i=1}^n$ is a basis of $\mathrm{V}_G^{\mathrm{PU}}$ as defined in (5) and define $\kappa(\mathrm{u}_G, \sigma(\mathrm{u}_G))$ by

$$\kappa(\mathrm{u}_G, \sigma(\mathrm{u}_G)) := \frac{\| M_T(\mathrm{u}_G) \|_{l^2}}{\| M_D(\mathrm{u}_G) \|_{l^2}}. \tag{14}$$

In contrast, Kim et al. propose [14] using Young's modulus divided by the characteristic length of a global finite element along the local boundary, so $\kappa \approx \frac{E}{h}$. In this case, κ scales with the material parameters and the global-level only.

Overall, the local problem with Robin boundary conditions then reads: Find $\mathrm{u}_L \in \mathrm{V}_L^{\mathrm{PU}} \subset H^1_{\Gamma_L \cap \Gamma_G^D}(\Omega_L)$ such that

$$\begin{aligned} &\int\limits_{\Omega_L} \sigma(\mathrm{u}_L) : \varepsilon(\mathrm{v}_L) \,\mathrm{d}x + \kappa \int\limits_{\Gamma_L \setminus \Gamma_G} \mathrm{u}_L \mathrm{v}_L \,\mathrm{d}s = \\ &\int\limits_{\Gamma_L \cap \Gamma_G^N} \bar{t}\, \mathrm{v}_L \,\mathrm{d}s + \int\limits_{\Gamma_L \setminus \Gamma_G} \Big(t(\mathrm{u}_G) + \kappa \mathrm{u}_G \Big) \mathrm{v}_L \,\mathrm{d}s + \int\limits_{\Omega_L} f\, \mathrm{v}_L \,\mathrm{d}x \end{aligned} \tag{15}$$

for all $\mathrm{v}_L \in \mathrm{V}_L^{\mathrm{PU}}$, where we typically set $\kappa := \kappa(\mathrm{u}_G, \sigma(\mathrm{u}_G))$ given in (14). Recall however, that the choice of $\kappa = 0$ yields Neumann boundary conditions and $\kappa \to \infty$ models Dirichlet boundary conditions.

Finally, we enrich the global space V_G^{PU} with the local solution u_L and solve the global problem (11) again, compare Fig. 1. As we solve the initial global problem without a computed enrichment, it provides inexact boundary data for the local problem. In order to obtain an improved enrichment, we can repeat the cycle, i.e. perform multiple global-local iterations.

With notation as in (4), the involved approximation spaces then evolve as follows: We solve the initial ($it = 0$) global problem without enrichments, i.e. we use

$$V_{G,0}^{PU} = \sum_i \varphi_i \mathcal{P}_i, \tag{16}$$

and all further global problems ($it > 0$) with the current local solution as an enrichment

$$V_{G,it}^{PU} = \sum_i \varphi_i \mathcal{P}_i + \sum_{i \in \mathcal{L}} \varphi_i \mathcal{E}_{it,gl} \quad (it > 0), \tag{17}$$

where L are the indices of global patches $\omega_{G,i}$ to be enriched with the local solution and the enrichment space $\mathcal{E}_{it,gl} = \text{span}\langle u_{L,it}\rangle$ is given by the current solution of the local problem. Through all iterations ($it > 0$) the local approximation space V_L^{PU} is given by

$$V_{L,it}^{PU} = \sum_i \varphi_i \mathcal{P}_i + \sum_{i \in C} \varphi_i \mathcal{E}_{crack} \tag{18}$$

where the set C collects the indices of patches on which we use the crack enrichments $\mathcal{E}_{crack}$ defined in e.g. [23]. Furthermore, the Robin parameter κ in (14) now also depends on the iteration κ_{it}, whereas in [14] it is fixed for all iterations (it). Note, that an equally valid approach is to construct the local problem around the crack tip only and consequently enrich all global problems with the Heaviside function too.

4 Numerical Results

For all our $2D$ experiments the global computational domain Ω_G is a cracked unit square. The crack tip is located at $(0.55, 0.34)$, from which the crack splits the square in $(-x_1)$-direction, compare Fig. 2. On this domain, we solve the equations of linear elasticity (7). Unless stated otherwise, we use Poisson's ratio $\nu = 0.3$, and Young's modulus $E = 10$ as fixed material parameters.

To define two model problems, we use terms of the expansion of a linear elasticity solution around a crack tip [5, 6]. Our first model problem is a simple mode I problem, given by the first term of the mode I expansion u_I^*. For our second model problem, we add the second term of the mode I expansion, the first term of the mode

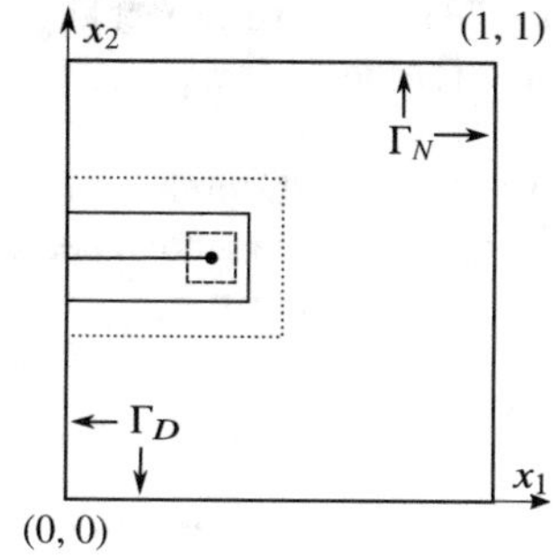

Fig. 2 The geometry of our $2D$ model problems: A cracked unit square. Dirichlet boundary conditions on the left and bottom boundary segments, Neumann boundary conditions on the top and left boundary segments. A thinner box around the crack indicates the local domain, the dotted box the local domain with a buffer zone. The dashed box is for computing errors around the tip

II expansion u_{II}^* and an oscillating cosine given by

$$osc(x) := \begin{pmatrix} \cos\left(8\left|x - x_{tip}\right|\right) \\ \cos\left(8\left|x - x_{tip}\right|\right) \end{pmatrix} \tag{19}$$

to define our analytic solution u_M^*. We refer to this second problem as the mixed-mode problem. We apply Dirichlet type boundary conditions on the bottom and left boundaries of the unit square Ω_G and Neumann type boundary conditions on the right and top boundary segments. We apply Neumann boundary conditions on the right and top boundary segments of the unit square Ω_G and on the bottom and left boundaries we apply Dirichlet boundary conditions.

In both model cases, we use a single local problem around the complete crack, where the local domain is the rectangle defined by the corners (0, 0.48125) and (0.39375, 0.64375), compare Fig. 2. As a consequence we have the true global boundary data on the left boundary segment. For each experiment we report the global and local discretization level l_g and l_l, as described in Sect. 2.1. Both global and local problems are solved with linear polynomials as $\mathcal{P}_i$ for all patches. Additionally, we use crack enrichments $\mathcal{E}_{crack}$ on the local problems. We enrich crack- but not tip-intersecting patches $\omega_{L,i}$ with a Heaviside function and patches $\omega_{L,i}$ around the tip with the first term of both the mode I and mode II expansion (u_I^* and u_{II}^*). Here we enrich geometrically, i.e., we enrich every patch whose midpoint is inside a box with edge length 0.06 around the tip. To prevent instabilities in the basis, we apply the stable transformation [24] once on the local problem, and in every iteration on the global problem, after enriching with the local solution. Before enriching with a new local solution, we erase the old enrichment on the global problem, compare (17). Note, that the crack on the global problem is only modeled by the computed enrichment. Therefore, we expect large errors on the initial global problems, as we actually solve for the wrong physics there, i.e. an uncracked unit square.

Throughout this paper we use two different norms to measure quantities of interest. The first being the L^2-norm and the second being the H^1-semi-norm, which for simplicity we refer to as just the H^1-norm

$$\|u\|_{L^2(\Omega_*)} = \left(\int_{\Omega_*} |u|^2 \, \mathrm{d}x\right)^{\frac{1}{2}} \quad \text{and} \quad |u|_{H^1(\Omega_*)} = \left(\int_{\Omega_*} \|\nabla u\|^2 \, \mathrm{d}x\right)^{\frac{1}{2}} \tag{20}$$

evaluated on some subset $\Omega_* \subset \Omega_G$ of the global computational domain Ω_G. When we report errors, we always compute the normalized global and local errors of iteration it with respect to the true solution, i.e. we define

$$e^{it}_{I/M,\Omega_*} := \frac{\left\| \mathrm{u}^*_{I/M} - \mathrm{u}^{it}_G \right\|_{\Omega_*}}{\left\| \mathrm{u}^*_{I/M} \right\|_{\Omega_*}} \tag{21}$$

on some subset $\Omega_* \subset \Omega_G$ of the global domain. We measure errors on the global domain Ω_G and on a small box with edge length 0.04 centered around the crack tip $\Omega_{tip} \subset \Omega_L$, see Fig. 2, to check if our approximations capture the near tip behavior correctly.

Example 1 For a first experiment we investigate the impact of the type of boundary conditions chosen on the local problem. Here we are also interested in the impact that the parameter κ in the Robin boundary conditions has on the overall accuracy. Figure 3 shows the measured global H^1-error (21) on the first and second global-

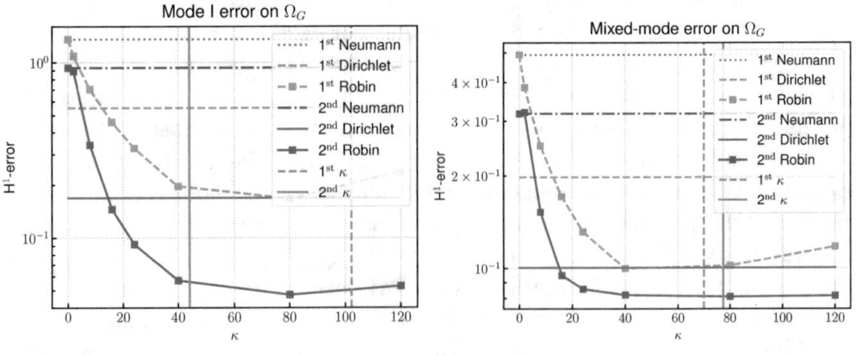

Fig. 3 Error on first and second iteration with Robin boundary conditions for various values of κ, Neumann ($\kappa = 0$) and Dirichlet (horizontal line) boundary conditions. Two vertical lines indicate the computed κ by (14). The problems are solved on $l_g = 5$ and $l_l = 4$

Algorithm 1: Global-local Iterations

Solve initial global problem ($it = 0$) for $u_{G,0}$;
for $it = 1$ **to** $max(it)$ **do**
  Solve local problem (15) for $u_{L,it}$ with boundary data $u_{G,it-1}$;
  Clear global enrichments and enrich V_G^{PU} with $u_{L,it}$;
  Solve enriched global problem (11) for $u_{G,it}$;

local iteration plotted against different values of κ. Horizontal lines show the results obtained for Dirichlet and Neumann boundary conditions. Furthermore we indicate the values of κ computed by our scheme (14) by vertical lines. We can observe that Robin outperform Dirichlet type boundary conditions in both iterations for appropriate κ. Only if κ is too small, which includes Neumann boundary conditions, they perform worse. As asserted, the optimal value of κ seems to depend on properties of the solution, since the mixed-mode problem requires higher values of κ. Obviously, our proposed scheme (14) for the automatic selection of κ yields close to optimal results in both iterations (actually in all iterations) and both model cases.

Example 2 Due to the approximation error of the computed global solution u_G, we apply inexact boundary data on the local boundary Γ_L, i.e. u_G instead of the true solution u^*. In this experiment we are interested in two techniques to overcome this limitation: Running multiple global-local iterations, see Algorithm 1 and adding a buffer zone to the local domain Ω_L. The question here is, whether we are able to compute an enrichment with the boundary data from the global problem comparable to one computed from the exact boundary data.

Using Caccioppoli's inequality Gupta et al. [13] derive a bound on the difference between the analytic solution of the local problem with exact boundary data u^{ex} and the finite dimensional approximation with inexact boundary data u_n^{inex}. With u^{inex} denoting the analytic solution of the local problem with inexact boundary data, we have

$$\left\| u^{ex} - u_n^{inex} \right\|_{E(\Omega_L^\delta)} \leq C \inf_{v \in V_n(\Omega_L)} \left\| u^{inex} - v \right\|_{E(\Omega_L)} + \frac{C_1}{\delta} \left\| u^{ex} - u^{inex} \right\|_{L^2(\Omega_L)}. \tag{22}$$

Here Ω_L is the local domain, $\Omega_L^\delta \subset \Omega_L$ a smaller subdomain and E denotes the energy norm. The scalar $\delta := \text{dist}\left(\Omega_L^\delta, \Omega_L\right) > 0$ is called the buffer zone size. So on Ω_L^δ, where we want to use the local solution as an enrichment, this error has two components: The discretization error of the local problem and the error in the boundary data, where the latter can be reduced by solving the global problem more accurately, but also by applying a buffer zone, i.e., computing the enrichment on a larger domain than it is actually used on. For the buffer zone in this experiment,

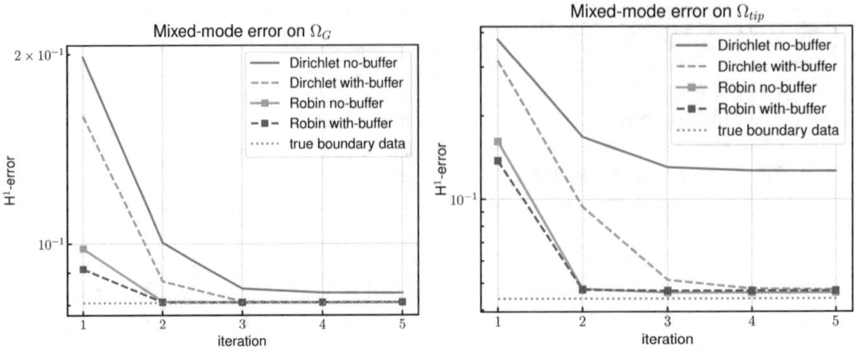

Fig. 4 Error over global-local iterations with and without buffer zone for Dirichlet and Robin boundary conditions (κ computed by (14)). In the buffer zone case we have $\delta = 0.1625$, which is 50% of the local domains width. A dotted line indicates a solution obtained with the true boundary data in Robin boundary conditions. The problems are solved with $l_g = 5$ and $l_l = 4$

we expand the local domain in all four directions by 50% of its width, hence $\delta = 0.1625$, compare Fig. 2. Besides increasing the global patch refinement level l_g, we can improve the boundary data by applying multiple global-local iterations. Here we reduce the error by improving the computed enrichment, until we are again limited by the applied discretization levels l_g and l_l. As in the previous experiment, we can observe that Robin outperform Dirichlet type boundary conditions, especially when we consider the error close to the tip, compare Fig. 4. They also require less iterations to reach an equilibrium. Applying a buffer zone, however, has minimal impact in combination with Robin boundary conditions. Yet a larger buffer zone allows Dirichlet boundary conditions to close the performance gap. It is remarkable that after some iterations we achieve accuracy comparable to applying the true solution of the global problem as local boundary data. We conclude that overcoming the problem of inexact boundary data requires multiple global-local iterations for both boundary condition types, whereas Dirichlet boundary conditions additionally require an enlarged local domain. Also, Robin type boundary conditions with the proposed iteration dependent parameter κ (14) yields the fastest convergence.

Example 3 As a last experiment on the cracked unit square, we compare our global-local scheme to a directly enriched global PUM applied to our mixed-mode problem. If the employed enrichments capture all non-smooth components of the solution, we expect to arrive at quadratic convergence (with respect to the patch size $diam(\omega_i)$) in L^2-norm and linear convergence in H^1-norm. In Fig. 5, we can observe that both methods yield exactly these rates. Besides that, the achieved accuracy is almost identical. Only near the tip, the global-local scheme performs better in the H^1-norm by a constant, most likely due to more accurately capturing the singularity in the derivative.

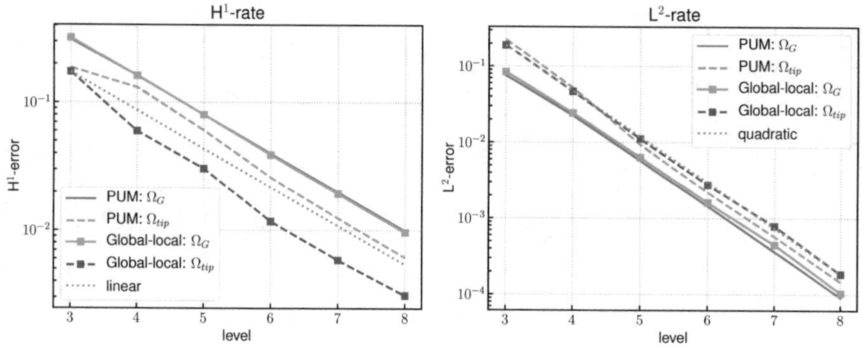

Fig. 5 Comparison of convergence rates in the mixed-mode problem obtained by a directly enriched global PUM and our global-local enrichment scheme. Dotted lines indicate optimal rates in the respective norms. Solved with $l_g = l_l$

Example 4 In order to test our PUM^{gl} on a more difficult problem, we assign two different materials to the unit square domain: We use two linear elastic materials with Poisson's ratio $\nu = 0.3$, where the upper half of the square has a Young's modulus of $E = 10$ and the lower half $E = 1$. A crack separates the materials from $(0.5, 0)$ to the tip at $(0.5, 0.5)$. Again we apply boundary conditions according to a known analytic solution of this type of problem, described in [29]. This problem is more difficult in the sense that we need 12 enrichments, also described in [29], to model the crack tip. However, these enrichments do not account for all effects due to the different materials, thus we do not expect them to completely resolve the crack tip singularity, compare the results in [29]. In addition to these singular enrichments, we enrich crack- but not tip-intersecting patches with a Heaviside function and patches at the interface of the materials with a ramp function. For our global-local scheme we apply the latter two directly on the global problem and construct the local problem to only cover a small area around the tip. Note that our scheme (14) to compute κ for Robin boundary conditions now has the advantage of respecting both material parameters along the local boundary without any changes. Figure 6 shows that, as expected, in this bimaterial problem both the PUM^{gl} and the enriched PUM do not attain quadratic and linear convergence in L^2- and H^1-norm respectively. Essentially by providing h-refinement towards the tip, the global-local enrichment scheme allows for lower errors by a constant in this setting.

Example 5 As a last example we compare the von Mises stress distribution in a 2D crack problem with either a single material or two material layers. To this end we apply the same boundary conditions in both problems: We fix the bottom and right edges of the unit square, i.e. apply Dirichlet zero boundary conditions and pull on the left edge below the crack to the left and on the upper edge to the top with unit force, i.e. apply Neumann boundary conditions. The solution is computed with three global-local iterations on global-level $l_g = 6$ and local-level $l_l = 5$, with

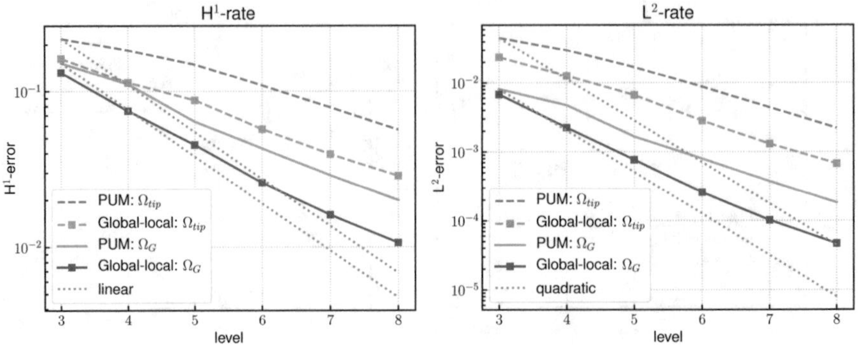

Fig. 6 Comparison of convergence rates in the bimaterial problem obtained by a global directly enriched PUM and our global-local enrichment scheme. Dotted lines indicate optimal rates in the respective norms. We and [29] do not achieve those, due to non-optimality of the employed enrichments. The local-level is the global-level

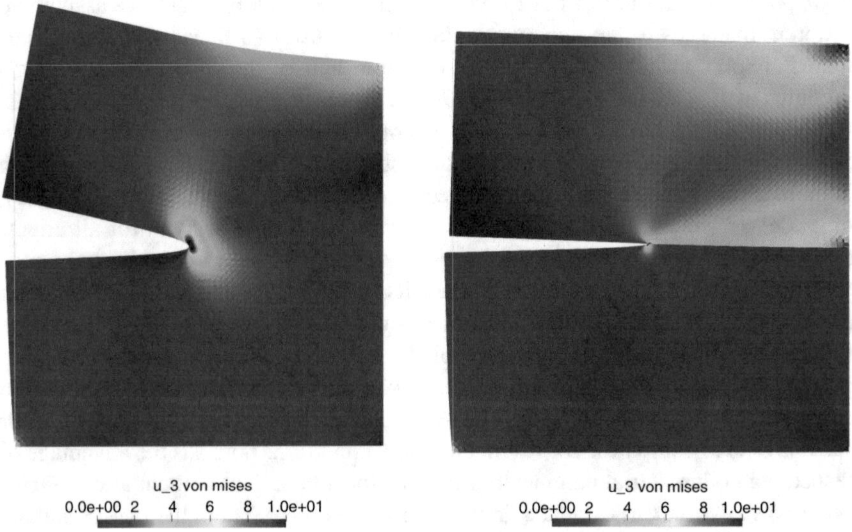

Fig. 7 Von Mises stress in single (left) and bimaterial (right) problem, as computed in the third iteration of our global-local enrichment scheme

additional h-refinement around the crack. On the local problem we again used Robin boundary conditions with κ computed by (14). Figures 7 and 8 show the von Mises stress distribution for this problem, once with a single material (left), where Young's modulus is globally $E = 1$ and on the right the bimaterial problem, where again the upper half is stiffer, with $E = 10$. Observe that the simulation nicely captures the increased stress along the material interface, as well as the difference in the stress distribution around the tip.

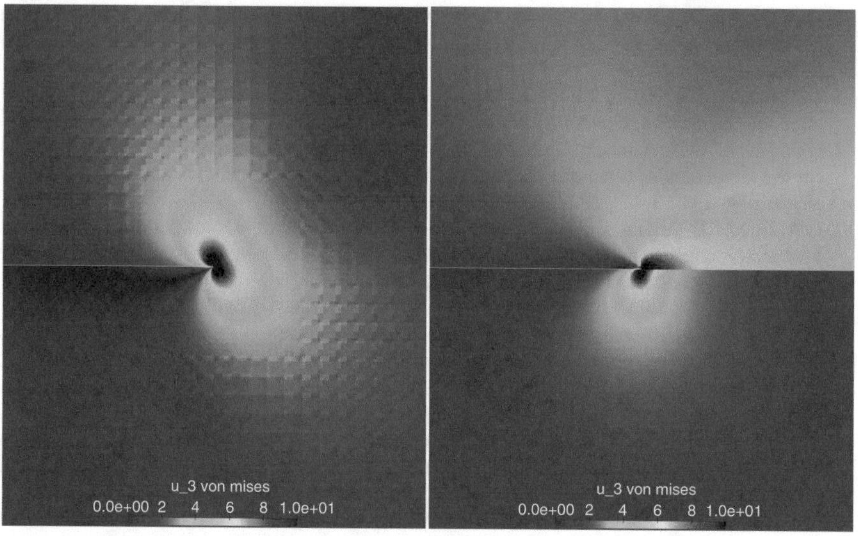

Fig. 8 Von Mises stress in single (left) and bimaterial (right) problem, as computed in the third iteration of our global-local enrichment scheme. Zoom to the tip

5 Concluding Remarks and Future Work

We have presented the realization of the global-local enrichment approach [7] for a general PUM and proposed an automatic scheme to identify the optimal parameter κ in Robin boundary conditions for each local solve in a global-local iteration. Our results show that the proposed scheme yields nearly optimal results and that Robin boundary conditions with this parameter κ do not require any buffer zone to attain fast convergence. We are currently working on efficient integration of the computed enrichments, together with parallelizing the proposed scheme. Moreover, we investigate re-using constant parts of the involved global and local linear systems, to further improve performance.

References

1. I. Babuška, R. Lipton, Optimal local approximation spaces for generalized finite element methods with application to multiscale problems. Multiscale Model. Simul. **9**(1), 373–406 (2011)
2. I. Babuška, X. Huang, R. Lipton, Machine computation using the exponentially convergent multiscale spectral generalized finite element method. ESAIM Math. Model Numer. Anal. **48**(2), 493–515 (2014)
3. D. Braess, *Finite Elemente: Theorie, schnelle Löser und Anwendungen in der Elastizitätstheorie* (Springer, Berlin, 2013)

4. C.A. Duarte, D.-J. Kim, Analysis and applications of a generalized finite element method with global-local enrichment functions. Comput. Methods Appl. Mech. Eng. **197**(6–8), 487–504 (2008)
5. C.A. Duarte, I. Babuška, J.T. Oden, Generalized finite element methods for three-dimensional structural mechanics problems. Comput. Struct. **77**(2), 215–232 (2000)
6. C.A. Duarte, O.N. Hamzeh, T.J. Liszka, W.W. Tworzydlo, A generalized finite element method for the simulation of three-dimensional dynamic crack propagation. Comput. Methods Appl. Mech. Eng. **190**(15–17), 2227–2262 (2001)
7. C.A. Duarte, D.-J. Kim, I. Babuška, A global-local approach for the construction of enrichment functions for the generalized FEM and its application to three-dimensional cracks, in *Advances in Meshfree Techniques* (Springer, Dordrecht, 2007), pp. 1–26
8. C.A. Felippa, *Introduction to Finite Element Methods* (University of Colorado, Boulder, 2001)
9. Fraunhofer SCAI, Puma - rapid enriched simulation application development, https://www.scai.fraunhofer.de/en/business-research-areas/meshfree-multiscale-methods/products/puma.html. Accessed 12 Dec 2018
10. M.J. Gander, F. Kwok, Best robin parameters for optimized schwarz methods at cross points. SIAM J. Sci. Comput. **34**(4), A1849–A1879 (2012)
11. M. Griebel, M.A. Schweitzer, A particle-partition of unity method for the solution of elliptic, parabolic and hyperbolic PDE. SIAM J. Sci. Comput. **22**(3), 853–890 (2000)
12. M. Griebel, M.A. Schweitzer, A particle-partition of unity method—part II: efficient cover construction and reliable integration. SIAM J. Sci. Comp. **23**(5), 1655–1682 (2002)
13. V. Gupta, D.-J. Kim, C.A. Duarte, Analysis and improvements of global-local enrichments for the generalized finite element method. Comput. Meth. Appl. Mech. Eng. **245**, 47–62 (2012)
14. D.-J. Kim, J.P. Pereira, C.A. Duarte, Analysis of three-dimensional fracture mechanics problems: a two-scale approach using coarse-generalized FEM meshes. Int. J. Numer. Methods Eng. **81**(3), 335–365 (2010)
15. D.-J. Kim, C.A. Duarte, N.A. Sobh, Parallel simulations of three-dimensional cracks using the generalized finite element method. Comput. Mech. **47**(3), 265–282 (2011)
16. J.M. Melenk, I. Babuška, The partition of unity finite element method: basic theory and applications. Comput. Meth. Appl. Mech. Eng. **139**(1–4), 289–314 (1996)
17. N. Moës, J. Dolbow, T. Belytschko, A finite element method for crack growth without remeshing. Int. J. Numer. Methods Eng. **46**(1), 131–150 (1999)
18. C.D. Mote, Global-local finite element. Int. J. Numer. Methods Eng. **3**(4), 565–574 (1971)
19. J.A. Plews, C.A. Duarte, A two-scale generalized finite element approach for modeling localized thermoplasticity. Int. J. Numer. Methods Eng. **108**(10), 1123–1158 (2016)
20. M. Schöllmann, M. Fulland, H.A. Richard, Development of a new software for adaptive crack growth simulations in 3d structures. Eng. Fract. Mech. **70**(2), 249–268 (2003)
21. M.A. Schweitzer, *A Parallel Multilevel Partition of Unity Method for Elliptic Partial Differential Equations*. Lecture Notes in Computational Science and Engineering, vol. 29 (Springer, Berlin, 2003)
22. M.A. Schweitzer, An algebraic treatment of essential boundary conditions in the particle–partition of unity method. SIAM J. Sci. Comput. **31**(2), 1581–1602 (2009)
23. M.A. Schweitzer, An adaptive hp-version of the multilevel particle–partition of unity method. Comput. Methods Appl. Mech. Eng. **198**, 1260–1272 (2009)
24. M.A. Schweitzer, Stable enrichment and local preconditioning in the particle–partition of unity method. Numer. Math. **118**(1), 137–170 (2011)
25. M.A. Schweitzer, Generalizations of the finite element method. Cent. Eur. J. Math. **10**, 3–24 (2012)
26. M.A. Schweitzer, S. Wu, Evaluation of local multiscale approximation spaces for partition of unity methods, in *Meshfree Methods for Partial Differential Equations VIII* (Springer, Cham, 2017), pp. 167–198
27. M.A. Schweitzer, A. Ziegenhagel, Rapid enriched simulation application development with puma, in *Scientific Computing and Algorithms in Industrial Simulations* (Springer, Cham, 2017), pp. 207–226

28. T. Strouboulis, K. Copps, I. Babuška, The generalized finite element method. Comput. Meth. Appl. Mech. Eng. **190**(32–33), 4081–4193 (2001)
29. N. Sukumar, Z.Y. Huang, J.-H. Prévost, Z. Suo, Partition of unity enrichment for bimaterial interface cracks. Int. J. Numer. Methods Eng. **59**(8), 1075–1102 (2004)
30. C.T. Sun, K.M. Mao, A global-local finite element method suitable for parallel computations. Comput. Struct. **29**(2), 309–315 (1988)
31. A. Ural, G. Heber, P.A. Wawrzynek, A.R. Ingraffea, D.G. Lewicki, J.B.C. Neto, Three-dimensional, parallel, finite element simulation of fatigue crack growth in a spiral bevel pinion gear. Eng. Fract. Mech. **72**(8), 1148–1170 (2005)
32. P.A. Wawrzynek, L.F. Martha, A.R. Ingraffea, A computational environment for the simulation of fracture processes in three dimensions, Analytical, numerical and experimental aspects of three dimensional fracture processes. ASME AMD **91**, 321–327 (1988)

Stable and Efficient Quantum Mechanical Calculations with PUMA on Triclinic Lattices

Clelia Albrecht, Constanze Klaar, and Marc Alexander Schweitzer

Abstract In this paper we are concerned with the efficient approximation of the Schrödinger eigenproblem using an orbital-enriched flat-top partition of unity method on general triclinic cells. To this end, we generalize the approach presented in Albrecht et al. (Comput. Meth. Appl. Mech. Eng. 342:224–239, 2018) via a simple yet effective transformation approach and discuss its realization in the PUMA software framework. The presented results clearly show that the proposed scheme attains all convergence and stability properties presented in Albrecht et al. (Comput. Meth. Appl. Mech. Eng. 342:224–239, 2018).

1 Introduction

The efficient treatment of the Schrödinger eigenproblem is an essential component of quantum mechanical material calculations. Enriched Galerkin methods can significantly reduce the number of degrees of freedom to attain the required chemical accuracy [1, 11]. While the partition of unity finite element method (PUFEM) of [11] provides this substantial reduction in the number of degrees of freedom and exponential convergence, it suffers from stability issues, i.e. ill-conditioning, due to the enrichment and requires the solution of a generalized eigenvalue problem which adversely effects the overall time to solution. The flat-top partition of unity method

C. Albrecht (✉) · C. Klaar
Fraunhofer-Institut für Algorithmen und Wissenschaftliches Rechnen SCAI, Sankt Augustin, Germany
e-mail: clelia.albrecht@scai.fraunhofer.de

M. A. Schweitzer
Fraunhofer-Institut für Algorithmen und Wissenschaftliches Rechnen SCAI, Sankt Augustin, Germany

Institut für Numerische Simulation, Rheinische Friedrich-Wilhelms-Universität Bonn, Bonn, Germany
e-mail: schweitzer@ins.uni-bonn.de

M. Griebel, M. A. Schweitzer (eds.), *Meshfree Methods for Partial Differential Equations IX*, Lecture Notes in Computational Science and Engineering 129,
https://doi.org/10.1007/978-3-030-15119-5_11

(FT-PUM) of [1] on the other hand overcomes both these issues while maintaining exponential convergence.

In this paper we generalize the approach presented in [1] to deal with arbitrary triclinic lattices. To this end, we transform the weak formulation of the Schrödinger eigenproblem on the triclinic cell back to the unit cube $[0, 1]^3$ and then apply the FT-PUM as derived in [1] for cuboid unit cells. This approach is easy to implement as well as maintaining the advantages the FT-PUM provides, namely the stability transformation and the variational mass lumping scheme, see [1] for details.

The remainder of this paper is structured as follows. First, we shortly review the FT-PUM and its properties in Sect. 2 before we present its application to the Schrödinger eigenproblem in Sect. 3. Then, we present the treatment of general triclinic lattices with the FT-PUM we have realized in PUMA [4], a general PUM software framework developed at Fraunhofer SCAI. The results of our numerical experiments are presented in Sect. 4 before we conclude with some remarks in Sect. 5.

2 Partition of Unity Method

In this Section, we briefly summarize only the aspects of the flat-top partition of unity method (FT-PUM) most important to its application to the Schrödinger eigenproblem. A more detailed description in this context can be found in [1], for more general details concerning the FT-PUM see [8]. All functionalities we describe here are implemented within the PUMA software framework developed at Fraunhofer SCAI [4].

The PUM is a generalization of classical finite element methods (FEM), developed to adapt this well-known spatial discretization technique for partial differential equations (PDE) to take into account problem-dependent a priori knowledge and thus overcome some disadvantages of classical FEM.

We start with the construction of a cover $C_\Omega := \{\omega_i\}$ of the computational domain Ω, by taking a uniformly refined mesh with mesh-width $2h$ and cells

$$C_i = \prod_{l=1}^{d} (o_i^l - h, o_i^l + h),$$

and rescale the cells to define overlapping patches ω_i of our cover as

$$\omega_i := \prod_{l=1}^{d} (o_i^l - \alpha h, o_i^l + \alpha h), \quad \text{with } \alpha \in (1, 2). \tag{1}$$

Note that, as $\alpha > 1$, the patches overlap, and with the choice of α, we can control the size of the overlap. Furthermore, because we also have $\alpha < 2$, on every patch there exists a subset $\widetilde{\omega}_i \subset \omega_i$ with nonvanishing measure that is only covered by this patch.

On these patches, we define weight functions $W_i : \Omega \to \mathbb{R}$ with $\mathrm{supp}(W_i) = \omega_i$ by

$$W_i(x) = \begin{cases} \mathcal{W} \circ T_i(x), & x \in \omega_i \\ 0, & \text{otherwise} \end{cases} \tag{2}$$

with the affine transforms $T_i : \overline{\omega}_i \to [-1, 1]^d$ and $\mathcal{W} : [-1, 1]^d \to \mathbb{R}$ any non-negative compactly supported function. By normalizing these weight functions we obtain the partition of unity functions

$$\varphi_i(x) := \frac{W_i(x)}{\sum_{l \in C_i} W_l(x)}, \tag{3}$$

where C_i denotes the neighborhood of ω_i. Due to this construction, we have $\varphi_i \equiv 1$ on the subset $\widetilde{\omega}_i$ of ω_i, i.e. a "flat top" [8]. These flat-top PU functions constitute the first ingredient of our FT-PUM. The second consists of local approximation spaces $V_i(\omega_i) := \mathrm{span}\langle \vartheta_i^m \rangle_{m=1}^{\dim(V_i)}$ defined on the patches ω_i. These approximation spaces, in general, consist of two parts: A smooth approximation space, for example polynomials $V_i^{\mathcal{P}}(\omega_i) := \mathrm{span}\langle \pi_i^s \rangle$, and a problem-dependent enrichment part $V_i^{\mathcal{E}}(\omega_i) := \mathrm{span}\langle \psi_i^t \rangle$. Thus, we can define the PUM space via

$$V^{\mathrm{PU}} := \sum_{i=1}^{N} \varphi_i V_i = \mathrm{span}\langle \varphi_i \vartheta_i^m \rangle; \tag{4}$$

that is, the basis functions of a PUM space are simply defined as the products of the PU functions φ_i and the local approximation functions ϑ_i^m, which, as the V_i consist of two parts, can be written as

$$\mathrm{span}\langle \vartheta_i^m \rangle = V_i(\omega_i) = V_i^{\mathcal{P}}(\omega_i) + V_i^{\mathcal{E}}(\omega_i) = \mathrm{span}\langle \pi_i^s, \psi_i^t \rangle. \tag{5}$$

Periodic boundary conditions are enforced by a slightly more general definition of patch neighborhoods C_i, see [1] for details. The global PUM space V^{PU} (4) attains its approximation properties essentially from the local spaces V_i. Thus, without any problem-dependent enrichment, i.e. $V_i^{\mathcal{E}} = \emptyset$, the PUM yields comparable error bounds as classical FEM [6, 8]. Yet, with appropriate enrichment spaces the PUM provides exponential convergence independent of the regularity of the solution, see e.g. [3, 9]. However, an arbitrary choice of $V_i^{\mathcal{E}}$ independently of $V_i^{\mathcal{P}}$ (and the PU φ_i) can lead to stability problems and highly ill-conditioned system matrices [2, 9, 11]. In our flat-top PUM, however, we can easily eliminate these stability issues without compromising the improved approximation quality due to enrichment with the help of a so-called stability transformation [9]. Thus, we only need to identify appropriate enrichment spaces for the considered application and attain a stable and highly accurate approximation with our flat-top PUM.

3 FT-PUM for the Schrödinger Eigenproblem

To this end, let us shortly review the model problem we consider in this paper, the one-electron Schrödinger equation in a periodic medium, which is represented by a parallelepiped unit cell $\Omega \subset \mathbb{R}^3$ with primitive lattice vectors $\boldsymbol{a}_d$ ($d = 1, 2, 3$). The strong form reads as

$$\begin{aligned} -\tfrac{1}{2}\nabla^2\psi(\boldsymbol{x}) + V_{\text{eff}}(\boldsymbol{x})\psi(\boldsymbol{x}) &= \varepsilon\psi(\boldsymbol{x}) && \text{in } \Omega, \\ \psi(\boldsymbol{x} + \boldsymbol{a}_d) &= \exp(i\boldsymbol{k}\cdot\boldsymbol{a}_d)\psi(\boldsymbol{x}) && \text{on } \Gamma_d, \\ \nabla\psi(\boldsymbol{x} + \boldsymbol{a}_d)\cdot\hat{\boldsymbol{n}}(\boldsymbol{x}) &= \exp(i\boldsymbol{k}\cdot\boldsymbol{a}_d)\nabla\psi(\boldsymbol{x})\cdot\hat{\boldsymbol{n}}(\boldsymbol{x}) && \text{on } \Gamma_d, \end{aligned} \tag{6}$$

where (ψ, ε) denotes an eigenpair consisting of the respective wavefunction ψ and its associated energy ε, $\hat{\boldsymbol{n}}(\boldsymbol{x})$ is the outward unit normal at $\boldsymbol{x}$ and Γ_d are the bounding faces of the domain Ω. Due to the periodicity of the medium the effective potential $V_{\text{eff}}(\boldsymbol{x})$ and the charge density ρ are periodic as well and satisfy

$$V_{\text{eff}}(\boldsymbol{x} + \boldsymbol{R}) = V_{\text{eff}}(\boldsymbol{x}), \quad \rho(\boldsymbol{x} + \boldsymbol{R}) = \rho(\boldsymbol{x}), \tag{7}$$

whereas the solution of Schrödinger's equation ψ, the so-called wavefunction, satisfies Bloch's theorem

$$\psi(\boldsymbol{x} + \boldsymbol{R}) = \psi(\boldsymbol{x})\exp(i\boldsymbol{k}\cdot\boldsymbol{R}) \tag{8}$$

for any lattice translation vector $\boldsymbol{R} = n_1\boldsymbol{a}_1 + n_2\boldsymbol{a}_2 + n_3\boldsymbol{a}_3$ with $n_d \in \mathbb{Z}$ ($d = 1, 2, 3$) and wavevector $\boldsymbol{k}$.

As in [11], to derive the weak form of (6), we test the one-electron Schrödinger equation with test functions $v \in \mathcal{V}$ with

$$\mathcal{V} := \{v \in H^1(\Omega, \mathbb{C}) : v(\boldsymbol{x} + \boldsymbol{a}_d) = v(\boldsymbol{x})\exp(i\boldsymbol{k}\cdot\boldsymbol{a}_d) \text{ on } \Gamma_d,\ d = 1, 2, 3\} \tag{9}$$

and integrate by parts to obtain

$$a(v, \psi) = \varepsilon\langle v, \psi\rangle_{L^2(\Omega,\mathbb{C})} \text{ for all } v \in \mathcal{V}, \tag{10}$$

where

$$a(v, \psi) := \frac{1}{2}\int_\Omega \left(\overline{\nabla v(\boldsymbol{x})}\nabla\psi(\boldsymbol{x}) + \overline{v(\boldsymbol{x})}V_{\text{eff}}(\boldsymbol{x})\psi(\boldsymbol{x})\right) d\boldsymbol{x} \tag{11}$$

and

$$\langle v, \psi\rangle_{L^2(\Omega,\mathbb{C})} := \int_\Omega \overline{v(\boldsymbol{x})}\psi(\boldsymbol{x})\, d\boldsymbol{x}. \tag{12}$$

Thus, we enforce the value-periodic condition as an essential boundary condition and the derivative-periodic condition as a natural one.

We use the flat-top PUM as described in Sect. 2 to construct a finite dimensional subspace $V_M \subset \mathcal{V}$ for the discretization of (10). To equip the discretization space V_M with potent approximation power we employ radial enrichment functions ψ_i^t constructed from isolated-atom solutions, see [1, 11] for details. Due to the use of our stability transform [9] the resulting discrete generalized eigenvalue problem

$$H\tilde{\psi} = \varepsilon S\tilde{\psi}, \tag{13}$$

where $H = (H_{ij}) \in \mathbb{C}^{M\times M}$ denotes the discrete Hamiltonian and $S = (S_{ij}) \in \mathbb{C}^{M\times M}$ is the so-called overlap (or consistent mass) matrix

$$H_{ij} := a(\phi_j, \phi_i), \text{ and } S_{ij} := \langle \phi_j, \phi_i \rangle_{L^2(\Omega,\mathbb{C})}, \tag{14}$$

involves well-conditioned matrices H and S and is thus in principle solvable. Yet, the solution of generalized eigenvalue problems is rather expensive compared with a standard eigenvalue, i.e. when $S = \mathbb{I}$. Fortunately, there exists a general variational mass lumping scheme [10] for the PUM that is applicable to arbitrary enriched approximations and preserves the improved approximation properties due to enrichment which allows us to transform (13) into a standard eigenvalue problem

$$H\tilde{u} = \varepsilon\tilde{u}, \tag{15}$$

see [1] for details.

3.1 Transformation on a Triclinic Unit Cell

To handle a triclinic rather than a cuboid unit cell geometry using the PUM as described in Sect. 2, we can pursue the following three approaches:

We can construct the triclinic unit cell geometry and cover it with axis-aligned patches to then carry out all integrations on the intersection of the geometry and the cover, cmp. Fig. 1. One of the problems with this procedure is, that the implementation of periodic boundary conditions is not as straight-forward as for axis-aligned domains presented in [1], because we can not guarantee in general that patches copied to the opposite boundary retain the flat-top property.

The second option, with a more involved implementation, is to construct the cover in a boundary-fitted way using triclinic cells, cmp. Fig. 1. In this case, we also integrate on general triclinic and not axis-aligned integration cells but the realization of periodic boundary conditions works analogously as for cuboid cells.

The third option, which was actually implemented for this paper, is to not use a triclinic geometry explicitly, but to transform the basis functions to the unit cube

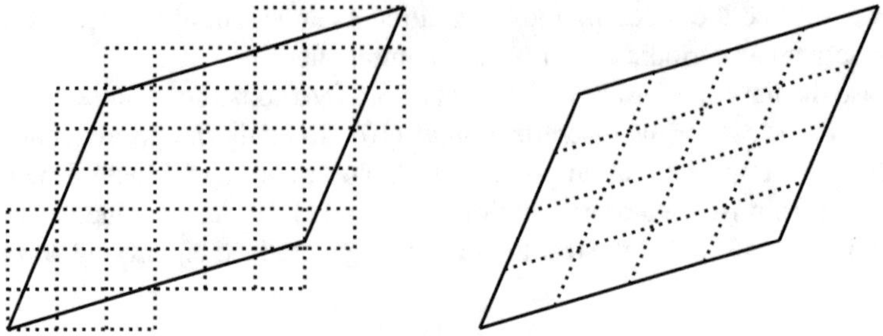

Fig. 1 Schematic two-dimensional representation of possible transformation approaches for triclinic cells not implemented for this paper. Left: Construct a cover of the triclinic cell with axis-aligned patches. Right: Construct a cover fitted to the triclinic cell

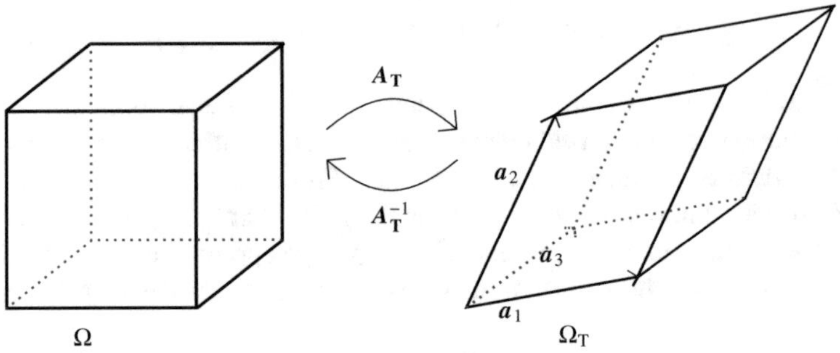

Fig. 2 Schematic representation of the transformation $\boldsymbol{A}_{\mathrm{T}} : \Omega \longrightarrow \Omega_{\mathrm{T}}$. This (linear) transformation can be represented by the matrix $\boldsymbol{A}_{\mathrm{T}} = (\boldsymbol{a}_1, \boldsymbol{a}_2, \boldsymbol{a}_3)$, whose column vectors are the primitive lattice vectors $\boldsymbol{a}_d$, $d = 1, 2, 3$ of Ω_{T}

$\Omega = [0, 1]^3$ using the transformation theorem and carry out all integration on there. This is the usual isoparametric approach also used in classical FEM approaches, see e.g. [7]. The transformation theorem reads as

$$\int_{\Phi(\Omega)} f(y)dy = \int_{\Omega} f(\Phi(x))|\mathrm{det} D\Phi(x)|dx \tag{16}$$

for a diffeomorphism $\Phi : \Omega \to \Phi(\Omega) = \Omega_{\mathrm{T}}$. For our case, the transformation maps the unit cube Ω to the triclinic unit cell Ω_{T}, i.e. it is the matrix $\boldsymbol{A}_{\mathrm{T}}$ whose column vectors are the primitive lattice vectors $\boldsymbol{a}_d$, $d = 1, 2, 3$ (see Fig. 2). The basis functions in PUMA are defined on a reference patch $\vartheta_i^m : \Omega \to \mathbb{R}$, i.e points have to be transformed from the triclinic unit cell to the unit cube via ${\boldsymbol{A}_{\mathrm{T}}}^{-1}$ in order to obtain the composite basis functions $\vartheta_i^m \circ {\boldsymbol{A}_{\mathrm{T}}}^{-1} : \Omega_{\mathrm{T}} \to \mathbb{R}$. In consequence, when applying the transformation theorem (16), both transformations cancel each other out, which leads to the transformed weak formulation

$$a_{\mathrm{T}}(v, \psi) = \varepsilon \langle v, \psi \rangle_{\mathrm{T},\, L^2(\Omega,\mathbb{C})} \text{ for all } v \in \mathcal{V}, \tag{17}$$

where

$$a_{\mathrm{T}}(v,\psi) := \frac{1}{2}\int_{\Omega}\Big(\overline{{A_{\mathrm{T}}}^{-T}\nabla v(\boldsymbol{x})}\,{A_{\mathrm{T}}}^{-T}\nabla\psi(\boldsymbol{x}) + \overline{v(\boldsymbol{x})}V_{\mathrm{eff}}(A_{\mathrm{T}}\boldsymbol{x})\psi(\boldsymbol{x})\Big)|\mathrm{det}A_{\mathrm{T}}|\,d\boldsymbol{x} \tag{18}$$

and

$$\langle v,\psi\rangle_{\mathrm{T},\,L^2(\Omega,\mathbb{C})} := \int_{\Omega}\overline{v(\boldsymbol{x})}\psi(\boldsymbol{x})|\mathrm{det}A_{\mathrm{T}}|\,d\boldsymbol{x}. \tag{19}$$

Note that this approach is remarkably easy to realize within the PUMA software framework where all functionalities as described in [1] are already implemented. We only have to change the weak formulation to solve the Schrödinger eigenproblem on a triclinic cell.

4 Numerical Results

In [1], we established that the flat-top PUM combined with radial enrichments for solving (10) works on cuboid unit cells subject to periodic and Bloch-periodic boundary conditions. Here, we concentrate on the numerical validation of the transformed problem (17) under Bloch-periodic boundary conditions. As stated in Sect. 2, we implemented this application of the flat-top PUM within the PUMA software framework developed at Fraunhofer SCAI [4]. PUMA is parallelized using MPI and for these experiments, we employed a simple tensor-product $6\times6\times6$ Gauß integration rule on subdivided cover cells [5, 8].

We consider a triclinic unit cell Ω_{T} with primitive lattice vectors $\boldsymbol{a}_1 := a(1, 0.02, -0.04)$, $\boldsymbol{a}_2 := a(0.06, 1.05, -0.08)$ and $\boldsymbol{a}_3 := a(0.10, -0.12, 1.10)$ and lattice parameter $a = 5$. For our benchmark problem on a transformed cell, we once again choose the periodic Gaussian potential, defined as

$$V(\boldsymbol{x}) = \sum_{\boldsymbol{R}} V_g(|\boldsymbol{x}-\boldsymbol{\tau}-\boldsymbol{R}|), \tag{20}$$

with

$$V_g(r) = -10\exp\left(-\frac{r^2}{2.25}\right), \tag{21}$$

where we sum over the lattice translation vectors $\boldsymbol{R} = i_1\boldsymbol{a}_1 + i_2\boldsymbol{a}_2 + i_3\boldsymbol{a}_3$ with $i_d = -2,\ldots,2\ (d = 1, 2, 3)$.

As described in Sect. 3.1, we transform this problem on the unit cube and solve the transformed weak formulation (17) subject to Bloch-periodic boundary conditions with $\boldsymbol{k} = (0.12, 0.23, 0.34)$ in reciprocal lattice coordinates. Thus this benchmark problem corresponds to the second benchmark problem discussed in [1] on a skewed lattice.

For our numerical experiments, we measure the absolute error with respect to a reference solution computed with classical cubic finite elements on a $64 \times 64 \times 64$ mesh, which is accurate to 7 digits (with $\lambda_1^{\text{ref}} = -5.9609433$ Ha and $\sum_{i=1}^{10} \lambda_i^{\text{ref}} = -26.3292418$ Ha). We first apply the flat-top PUM without any enrichments and linear, quadratic and cubic polynomial spaces on uniformly refined covers. As for cuboid unit cells, we expect convergence rates comparable to classical FEM [1]. For the second experiment, we fix a uniformly refined cover and increase the enrichment support radius for a single enrichment function, i.e. we enrich a patch on the unit cube if its midpoint lies within a certain distance to the potential center on the triclinic cell. The results of these experiments are shown in Fig. 3, where we can indeed observe the anticipated convergence rates for the not-enriched case and spectral convergence in the enriched case for the lowest state. As in [1], we see that the benefits of enriching are striking: The typically desired accuracy of 10^{-3} Ha is achieved with just 135 for the linear enriched approximation on a $3 \times 3 \times 3$ cover. Compared to that, we do not even achieve chemical accuracy when using an approximation approach involving only linear polynomials with more that 10^7 DOFs, while the best polynomial approximation still needs 81,920 DOFs.

Furthermore, we examine the influence of the stretch factor α (1) on the accuracy and convergence of the lumped eigenvalue problem. To this end, we use ten enrichment functions and observe the convergence of the ten lowest states for $\alpha = 1.1, 1.3, 1.5$ (see Fig. 4). In this case, contrary to the observations we made in [1] for the harmonic oscillator benchmark problem, the stretch factor does not have a real influence on the monotonicity of the convergence of this more complex problem.

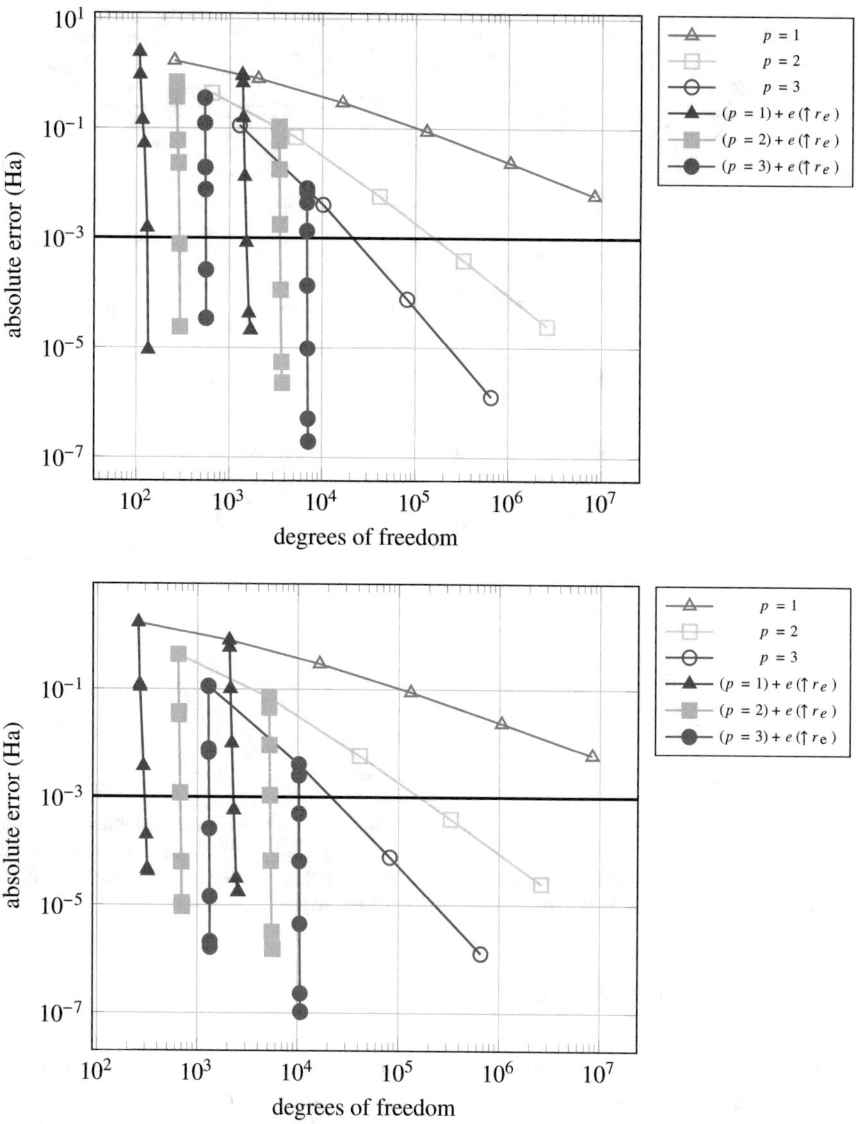

Fig. 3 Convergence history of the lowest eigenvalue λ_1 for the gaussian oscillator potential attained for different refinement schemes. We consider a purely polynomial approximation ($p = 1, 2, 3$) on a sequence of uniformly refined covers, which shows the expected $2p$-convergence rates (see [1]). Furthermore, we consider a refinement by increasing the enrichment radius with a single enrichment function on a fixed uniform cover (top: $3 \times 3 \times 3$, $7 \times 7 \times 7$; bottom: $4 \times 4 \times 4$, $8 \times 8 \times 8$) that is labeled by $p = 1, 2, 3 + e \uparrow r_e$, where we observe spectral convergence

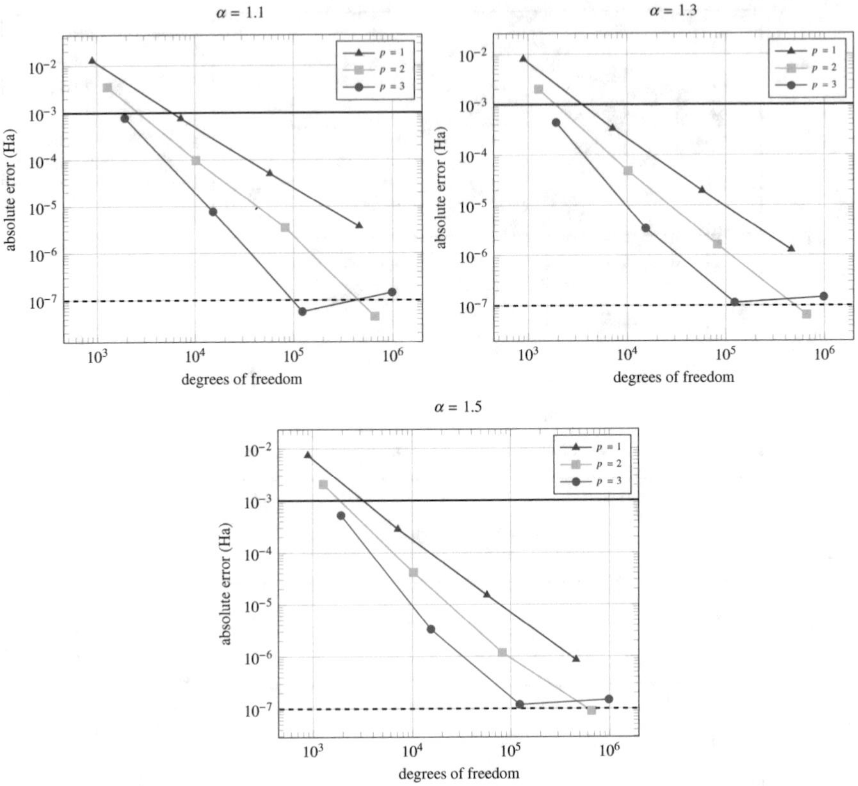

Fig. 4 Convergence history of the sum of the ten lowest eigenvalues for the harmonic oscillator potential obtained for different values of $\alpha = 1.1, 1.3, 1.5$ (left to right) on uniformly refined covers with ten enrichment functions employed on every patch ω_i with a lumped overlap matrix. The dashed line indicates the accuracy of the employed reference solution

5 Concluding Remarks

In this paper, we extended the implementation of the orbital-enriched flat-top PUM for the Schrödinger eigenproblem to work on arbitrary triclinic cells. As for cubic cells, we observe that our approach yields a stable and efficient way to approximate the solution. Incorporating enrichment functions into the local approximation drastically reduces the number of DOFs needed to achieve chemical accuracy. This more complex problem does not show the same dependence on the stretch factor α as we have observed for other benchmark problems—a good a priori method to determine the maximum α for a given problem remains an open question.

We are currently working on the implementation of the full Kohn–Sham loop and on the incorporation of nonlocal pseudopotentials. Another interesting question would be how the transformation approach we used here compares to different geometric transformation approaches we can realize within PUMA.

References

1. C. Albrecht, C. Klaar, J.E. Pask, M.A. Schweitzer, N. Sukumar, A. Ziegenhagel, Orbital-enriched flat-top partition of unity method for the schrödinger eigenproblem. Comput. Meth. Appl. Mech. Eng. **342**, 224–239 (2018)
2. I. Babuška, U. Banerjee, Stable generalized finite element method (SGFEM). Comput. Meth. Appl. Mech. Eng. **201–204**(Suppl. C), 91–111 (2012)
3. I. Babuska, R. Lipton, Optimal local approximation spaces for generalized finite element methods with application to multiscale problems. Multiscale Model. Simul. **9**(1), 373–406 (2011)
4. Fraunhofer SCAI, *Puma - Rapid Enriched Simulation Application Development*, https://www.scai.fraunhofer.de/en/business-research-areas/meshfree-multiscale-methods/products/puma.html. Accessed 12 Dec 2018
5. M. Griebel, M.A. Schweitzer, A particle-partition of unity method—part II: efficient cover construction and reliable integration. SIAM J. Sci. Comput. **23**(5), 1655–1682 (2002)
6. J.M. Melenk, I. Babuška, The partition of unity finite element method: Basic theory and applications. Comput. Meth. Appl. Mech. Eng. **139**, 289–314 (1996)
7. J.E. Pask, B.M. Klein, P.A. Sterne, C.Y. Fong, Finite-element methods in electronic-structure theory. Comput. Phys. Commun. **135**(1), 1–34 (2001)
8. M.A. Schweitzer, A parallel multilevel partition of unity method for elliptic partial differential equations. *Lecture Notes in Computational Science and Engineering*, vol. 29 (Springer, Cham, 2003)
9. M.A. Schweitzer, Stable enrichment and local preconditioning in the particle–partition of unity method. Numer. Math. **118**(1), 137–170 (2011)
10. M.A. Schweitzer, Variational mass lumping in the partition of unity method. SIAM J. Sci. Comput. **35**(2), A1073–A1097 (2013)
11. N. Sukumar, J.E. Pask, Classical and enriched finite element formulations for Bloch-periodic boundary conditions. Int. J. Numer. Methods Eng. **77**(8), 1121–1138 (2009)

Editorial Policy

1. Volumes in the following three categories will be published in LNCSE:

i) Research monographs
ii) Tutorials
iii) Conference proceedings

Those considering a book which might be suitable for the series are strongly advised to contact the publisher or the series editors at an early stage.

2. Categories i) and ii). Tutorials are lecture notes typically arising via summer schools or similar events, which are used to teach graduate students. These categories will be emphasized by Lecture Notes in Computational Science and Engineering. **Submissions by interdisciplinary teams of authors are encouraged**. The goal is to report new developments – quickly, informally, and in a way that will make them accessible to non-specialists. In the evaluation of submissions timeliness of the work is an important criterion. Texts should be well-rounded, well-written and reasonably self-contained. In most cases the work will contain results of others as well as those of the author(s). In each case the author(s) should provide sufficient motivation, examples, and applications. In this respect, Ph.D. theses will usually be deemed unsuitable for the Lecture Notes series. Proposals for volumes in these categories should be submitted either to one of the series editors or to Springer-Verlag, Heidelberg, and will be refereed. A provisional judgement on the acceptability of a project can be based on partial information about the work: a detailed outline describing the contents of each chapter, the estimated length, a bibliography, and one or two sample chapters – or a first draft. A final decision whether to accept will rest on an evaluation of the completed work which should include

- at least 100 pages of text;
- a table of contents;
- an informative introduction perhaps with some historical remarks which should be accessible to readers unfamiliar with the topic treated;
- a subject index.

3. Category iii). Conference proceedings will be considered for publication provided that they are both of exceptional interest and devoted to a single topic. One (or more) expert participants will act as the scientific editor(s) of the volume. They select the papers which are suitable for inclusion and have them individually refereed as for a journal. Papers not closely related to the central topic are to be excluded. Organizers should contact the Editor for CSE at Springer at the planning stage, see *Addresses* below.

In exceptional cases some other multi-author-volumes may be considered in this category.

4. Only works in English will be considered. For evaluation purposes, manuscripts may be submitted in print or electronic form, in the latter case, preferably as pdf- or zipped ps-files. Authors are requested to use the LaTeX style files available from Springer at http://www.springer.com/gp/authors-editors/book-authors-editors/manuscript-preparation/5636 (Click on LaTeX Template → monographs or contributed books).

For categories ii) and iii) we strongly recommend that all contributions in a volume be written in the same LaTeX version, preferably LaTeX2e. Electronic material can be included if appropriate. Please contact the publisher.

Careful preparation of the manuscripts will help keep production time short besides ensuring satisfactory appearance of the finished book in print and online.

5. The following terms and conditions hold. Categories i), ii) and iii):

Authors receive 50 free copies of their book. No royalty is paid.
Volume editors receive a total of 50 free copies of their volume to be shared with authors, but no royalties.

Authors and volume editors are entitled to a discount of 40 % on the price of Springer books purchased for their personal use, if ordering directly from Springer.

6. Springer secures the copyright for each volume.

Addresses:

Timothy J. Barth
NASA Ames Research Center
NAS Division
Moffett Field, CA 94035, USA
barth@nas.nasa.gov

Michael Griebel
Institut für Numerische Simulation
der Universität Bonn
Wegelerstr. 6
53115 Bonn, Germany
griebel@ins.uni-bonn.de

David E. Keyes
Mathematical and Computer Sciences
and Engineering
King Abdullah University of Science
and Technology
P.O. Box 55455
Jeddah 21534, Saudi Arabia
david.keyes@kaust.edu.sa

and

Department of Applied Physics
and Applied Mathematics
Columbia University
500 W. 120 th Street
New York, NY 10027, USA
kd2112@columbia.edu

Risto M. Nieminen
Department of Applied Physics
Aalto University School of Science
and Technology
00076 Aalto, Finland
risto.nieminen@aalto.fi

Dirk Roose
Department of Computer Science
Katholieke Universiteit Leuven
Celestijnenlaan 200A
3001 Leuven-Heverlee, Belgium
dirk.roose@cs.kuleuven.be

Tamar Schlick
Department of Chemistry
and Courant Institute
of Mathematical Sciences
New York University
251 Mercer Street
New York, NY 10012, USA
schlick@nyu.edu

Editor for Computational Science
and Engineering at Springer:

Martin Peters
Springer-Verlag
Mathematics Editorial IV
Tiergartenstrasse 17
69121 Heidelberg, Germany
martin.peters@springer.com

Lecture Notes in Computational Science and Engineering

1. D. Funaro, *Spectral Elements for Transport-Dominated Equations.*
2. H.P. Langtangen, *Computational Partial Differential Equations.* Numerical Methods and Diffpack Programming.
3. W. Hackbusch, G. Wittum (eds.), *Multigrid Methods V.*
4. P. Deuflhard, J. Hermans, B. Leimkuhler, A.E. Mark, S. Reich, R.D. Skeel (eds.), *Computational Molecular Dynamics: Challenges, Methods, Ideas.*
5. D. Kröner, M. Ohlberger, C. Rohde (eds.), *An Introduction to Recent Developments in Theory and Numerics for Conservation Laws.*
6. S. Turek, *Efficient Solvers for Incompressible Flow Problems.* An Algorithmic and Computational Approach.
7. R. von Schwerin, ***M**ulti **B**ody **S**ystem **SIM**ulation.* Numerical Methods, Algorithms, and Software.
8. H.-J. Bungartz, F. Durst, C. Zenger (eds.), *High Performance Scientific and Engineering Computing.*
9. T.J. Barth, H. Deconinck (eds.), *High-Order Methods for Computational Physics.*
10. H.P. Langtangen, A.M. Bruaset, E. Quak (eds.), *Advances in Software Tools for Scientific Computing.*
11. B. Cockburn, G.E. Karniadakis, C.-W. Shu (eds.), *Discontinuous Galerkin Methods.* Theory, Computation and Applications.
12. U. van Rienen, *Numerical Methods in Computational Electrodynamics.* Linear Systems in Practical Applications.
13. B. Engquist, L. Johnsson, M. Hammill, F. Short (eds.), *Simulation and Visualization on the Grid.*
14. E. Dick, K. Riemslagh, J. Vierendeels (eds.), *Multigrid Methods VI.*
15. A. Frommer, T. Lippert, B. Medeke, K. Schilling (eds.), *Numerical Challenges in Lattice Quantum Chromodynamics.*
16. J. Lang, *Adaptive Multilevel Solution of Nonlinear Parabolic PDE Systems.* Theory, Algorithm, and Applications.
17. B.I. Wohlmuth, *Discretization Methods and Iterative Solvers Based on Domain Decomposition.*
18. U. van Rienen, M. Günther, D. Hecht (eds.), *Scientific Computing in Electrical Engineering.*
19. I. Babuška, P.G. Ciarlet, T. Miyoshi (eds.), *Mathematical Modeling and Numerical Simulation in Continuum Mechanics.*
20. T.J. Barth, T. Chan, R. Haimes (eds.), *Multiscale and Multiresolution Methods.* Theory and Applications.
21. M. Breuer, F. Durst, C. Zenger (eds.), *High Performance Scientific and Engineering Computing.*
22. K. Urban, *Wavelets in Numerical Simulation.* Problem Adapted Construction and Applications.
23. L.F. Pavarino, A. Toselli (eds.), *Recent Developments in Domain Decomposition Methods.*

24. T. Schlick, H.H. Gan (eds.), *Computational Methods for Macromolecules: Challenges and Applications.*

25. T.J. Barth, H. Deconinck (eds.), *Error Estimation and Adaptive Discretization Methods in Computational Fluid Dynamics.*

26. M. Griebel, M.A. Schweitzer (eds.), *Meshfree Methods for Partial Differential Equations.*

27. S. Müller, *Adaptive Multiscale Schemes for Conservation Laws.*

28. C. Carstensen, S. Funken, W. Hackbusch, R.H.W. Hoppe, P. Monk (eds.), *Computational Electromagnetics.*

29. M.A. Schweitzer, *A Parallel Multilevel Partition of Unity Method for Elliptic Partial Differential Equations.*

30. T. Biegler, O. Ghattas, M. Heinkenschloss, B. van Bloemen Waanders (eds.), *Large-Scale PDE-Constrained Optimization.*

31. M. Ainsworth, P. Davies, D. Duncan, P. Martin, B. Rynne (eds.), *Topics in Computational Wave Propagation.* Direct and Inverse Problems.

32. H. Emmerich, B. Nestler, M. Schreckenberg (eds.), *Interface and Transport Dynamics.* Computational Modelling.

33. H.P. Langtangen, A. Tveito (eds.), *Advanced Topics in Computational Partial Differential Equations.* Numerical Methods and Diffpack Programming.

34. V. John, *Large Eddy Simulation of Turbulent Incompressible Flows.* Analytical and Numerical Results for a Class of LES Models.

35. E. Bänsch (ed.), *Challenges in Scientific Computing - CISC 2002.*

36. B.N. Khoromskij, G. Wittum, *Numerical Solution of Elliptic Differential Equations by Reduction to the Interface.*

37. A. Iske, *Multiresolution Methods in Scattered Data Modelling.*

38. S.-I. Niculescu, K. Gu (eds.), *Advances in Time-Delay Systems.*

39. S. Attinger, P. Koumoutsakos (eds.), *Multiscale Modelling and Simulation.*

40. R. Kornhuber, R. Hoppe, J. Périaux, O. Pironneau, O. Wildlund, J. Xu (eds.), *Domain Decomposition Methods in Science and Engineering.*

41. T. Plewa, T. Linde, V.G. Weirs (eds.), *Adaptive Mesh Refinement – Theory and Applications.*

42. A. Schmidt, K.G. Siebert, *Design of Adaptive Finite Element Software.* The Finite Element Toolbox ALBERTA.

43. M. Griebel, M.A. Schweitzer (eds.), *Meshfree Methods for Partial Differential Equations II.*

44. B. Engquist, P. Lötstedt, O. Runborg (eds.), *Multiscale Methods in Science and Engineering.*

45. P. Benner, V. Mehrmann, D.C. Sorensen (eds.), *Dimension Reduction of Large-Scale Systems.*

46. D. Kressner, *Numerical Methods for General and Structured Eigenvalue Problems.*

47. A. Boriçi, A. Frommer, B. Joó, A. Kennedy, B. Pendleton (eds.), *QCD and Numerical Analysis III.*

48. F. Graziani (ed.), *Computational Methods in Transport.*

49. B. Leimkuhler, C. Chipot, R. Elber, A. Laaksonen, A. Mark, T. Schlick, C. Schütte, R. Skeel (eds.), *New Algorithms for Macromolecular Simulation.*

50. M. Bücker, G. Corliss, P. Hovland, U. Naumann, B. Norris (eds.), *Automatic Differentiation: Applications, Theory, and Implementations.*

51. A.M. Bruaset, A. Tveito (eds.), *Numerical Solution of Partial Differential Equations on Parallel Computers.*

52. K.H. Hoffmann, A. Meyer (eds.), *Parallel Algorithms and Cluster Computing.*

53. H.-J. Bungartz, M. Schäfer (eds.), *Fluid-Structure Interaction.*

54. J. Behrens, *Adaptive Atmospheric Modeling.*

55. O. Widlund, D. Keyes (eds.), *Domain Decomposition Methods in Science and Engineering XVI.*

56. S. Kassinos, C. Langer, G. Iaccarino, P. Moin (eds.), *Complex Effects in Large Eddy Simulations.*

57. M. Griebel, M.A Schweitzer (eds.), *Meshfree Methods for Partial Differential Equations III.*

58. A.N. Gorban, B. Kégl, D.C. Wunsch, A. Zinovyev (eds.), *Principal Manifolds for Data Visualization and Dimension Reduction.*

59. H. Ammari (ed.), *Modeling and Computations in Electromagnetics: A Volume Dedicated to Jean-Claude Nédélec.*

60. U. Langer, M. Discacciati, D. Keyes, O. Widlund, W. Zulehner (eds.), *Domain Decomposition Methods in Science and Engineering XVII.*

61. T. Mathew, *Domain Decomposition Methods for the Numerical Solution of Partial Differential Equations.*

62. F. Graziani (ed.), *Computational Methods in Transport: Verification and Validation.*

63. M. Bebendorf, *Hierarchical Matrices.* A Means to Efficiently Solve Elliptic Boundary Value Problems.

64. C.H. Bischof, H.M. Bücker, P. Hovland, U. Naumann, J. Utke (eds.), *Advances in Automatic Differentiation.*

65. M. Griebel, M.A. Schweitzer (eds.), *Meshfree Methods for Partial Differential Equations IV.*

66. B. Engquist, P. Lötstedt, O. Runborg (eds.), *Multiscale Modeling and Simulation in Science.*

67. I.H. Tuncer, Ü. Gülcat, D.R. Emerson, K. Matsuno (eds.), *Parallel Computational Fluid Dynamics 2007.*

68. S. Yip, T. Diaz de la Rubia (eds.), *Scientific Modeling and Simulations.*

69. A. Hegarty, N. Kopteva, E. O'Riordan, M. Stynes (eds.), *BAIL* 2008 – *Boundary and Interior Layers.*

70. M. Bercovier, M.J. Gander, R. Kornhuber, O. Widlund (eds.), *Domain Decomposition Methods in Science and Engineering XVIII.*

71. B. Koren, C. Vuik (eds.), *Advanced Computational Methods in Science and Engineering.*

72. M. Peters (ed.), *Computational Fluid Dynamics for Sport Simulation.*

73. H.-J. Bungartz, M. Mehl, M. Schäfer (eds.), *Fluid Structure Interaction II - Modelling, Simulation, Optimization.*

74. D. Tromeur-Dervout, G. Brenner, D.R. Emerson, J. Erhel (eds.), *Parallel Computational Fluid Dynamics 2008.*

75. A.N. Gorban, D. Roose (eds.), *Coping with Complexity: Model Reduction and Data Analysis.*

76. J.S. Hesthaven, E.M. Rønquist (eds.), *Spectral and High Order Methods for Partial Differential Equations.*

77. M. Holtz, *Sparse Grid Quadrature in High Dimensions with Applications in Finance and Insurance.*

78. Y. Huang, R. Kornhuber, O.Widlund, J. Xu (eds.), *Domain Decomposition Methods in Science and Engineering XIX.*

79. M. Griebel, M.A. Schweitzer (eds.), *Meshfree Methods for Partial Differential Equations V.*

80. P.H. Lauritzen, C. Jablonowski, M.A. Taylor, R.D. Nair (eds.), *Numerical Techniques for Global Atmospheric Models.*

81. C. Clavero, J.L. Gracia, F.J. Lisbona (eds.), *BAIL 2010 – Boundary and Interior Layers, Computational and Asymptotic Methods.*

82. B. Engquist, O. Runborg, Y.R. Tsai (eds.), *Numerical Analysis and Multiscale Computations.*

83. I.G. Graham, T.Y. Hou, O. Lakkis, R. Scheichl (eds.), *Numerical Analysis of Multiscale Problems.*

84. A. Logg, K.-A. Mardal, G. Wells (eds.), *Automated Solution of Differential Equations by the Finite Element Method.*

85. J. Blowey, M. Jensen (eds.), *Frontiers in Numerical Analysis - Durham 2010.*

86. O. Kolditz, U.-J. Gorke, H. Shao, W. Wang (eds.), *Thermo-Hydro-Mechanical-Chemical Processes in Fractured Porous Media - Benchmarks and Examples.*

87. S. Forth, P. Hovland, E. Phipps, J. Utke, A. Walther (eds.), *Recent Advances in Algorithmic Differentiation.*

88. J. Garcke, M. Griebel (eds.), *Sparse Grids and Applications.*

89. M. Griebel, M.A. Schweitzer (eds.), *Meshfree Methods for Partial Differential Equations VI.*

90. C. Pechstein, *Finite and Boundary Element Tearing and Interconnecting Solvers for Multiscale Problems.*

91. R. Bank, M. Holst, O. Widlund, J. Xu (eds.), *Domain Decomposition Methods in Science and Engineering XX.*

92. H. Bijl, D. Lucor, S. Mishra, C. Schwab (eds.), *Uncertainty Quantification in Computational Fluid Dynamics.*

93. M. Bader, H.-J. Bungartz, T. Weinzierl (eds.), *Advanced Computing.*

94. M. Ehrhardt, T. Koprucki (eds.), *Advanced Mathematical Models and Numerical Techniques for Multi-Band Effective Mass Approximations.*

95. M. Azaïez, H. El Fekih, J.S. Hesthaven (eds.), *Spectral and High Order Methods for Partial Differential Equations ICOSAHOM 2012.*

96. F. Graziani, M.P. Desjarlais, R. Redmer, S.B. Trickey (eds.), *Frontiers and Challenges in Warm Dense Matter.*

97. J. Garcke, D. Pflüger (eds.), *Sparse Grids and Applications – Munich 2012.*

98. J. Erhel, M. Gander, L. Halpern, G. Pichot, T. Sassi, O. Widlund (eds.), *Domain Decomposition Methods in Science and Engineering XXI.*

99. R. Abgrall, H. Beaugendre, P.M. Congedo, C. Dobrzynski, V. Perrier, M. Ricchiuto (eds.), *High Order Nonlinear Numerical Methods for Evolutionary PDEs - HONOM 2013.*

100. M. Griebel, M.A. Schweitzer (eds.), *Meshfree Methods for Partial Differential Equations VII.*

101. R. Hoppe (ed.), *Optimization with PDE Constraints - OPTPDE 2014.*

102. S. Dahlke, W. Dahmen, M. Griebel, W. Hackbusch, K. Ritter, R. Schneider, C. Schwab, H. Yserentant (eds.), *Extraction of Quantifiable Information from Complex Systems.*

103. A. Abdulle, S. Deparis, D. Kressner, F. Nobile, M. Picasso (eds.), *Numerical Mathematics and Advanced Applications - ENUMATH 2013.*

104. T. Dickopf, M.J. Gander, L. Halpern, R. Krause, L.F. Pavarino (eds.), *Domain Decomposition Methods in Science and Engineering XXII.*

105. M. Mehl, M. Bischoff, M. Schäfer (eds.), *Recent Trends in Computational Engineering - CE2014.* Optimization, Uncertainty, Parallel Algorithms, Coupled and Complex Problems.

106. R.M. Kirby, M. Berzins, J.S. Hesthaven (eds.), *Spectral and High Order Methods for Partial Differential Equations - ICOSAHOM'14.*

107. B. Jüttler, B. Simeon (eds.), *Isogeometric Analysis and Applications 2014.*

108. P. Knobloch (ed.), *Boundary and Interior Layers, Computational and Asymptotic Methods – BAIL 2014.*

109. J. Garcke, D. Pflüger (eds.), *Sparse Grids and Applications – Stuttgart 2014.*

110. H. P. Langtangen, *Finite Difference Computing with Exponential Decay Models.*

111. A. Tveito, G.T. Lines, *Computing Characterizations of Drugs for Ion Channels and Receptors Using Markov Models.*

112. B. Karazösen, M. Manguoğlu, M. Tezer-Sezgin, S. Göktepe, Ö. Uğur (eds.), *Numerical Mathematics and Advanced Applications - ENUMATH 2015.*

113. H.-J. Bungartz, P. Neumann, W.E. Nagel (eds.), *Software for Exascale Computing - SPPEXA 2013-2015.*

114. G.R. Barrenechea, F. Brezzi, A. Cangiani, E.H. Georgoulis (eds.), *Building Bridges: Connections and Challenges in Modern Approaches to Numerical Partial Differential Equations.*

115. M. Griebel, M.A. Schweitzer (eds.), *Meshfree Methods for Partial Differential Equations VIII.*

116. C.-O. Lee, X.-C. Cai, D.E. Keyes, H.H. Kim, A. Klawonn, E.-J. Park, O.B. Widlund (eds.), *Domain Decomposition Methods in Science and Engineering XXIII.*

117. T. Sakurai, S.-L. Zhang, T. Imamura, Y. Yamamoto, Y. Kuramashi, T. Hoshi (eds.), *Eigenvalue Problems: Algorithms, Software and Applications in Petascale Computing.* EPASA 2015, Tsukuba, Japan, September 2015.

118. T. Richter (ed.), *Fluid-structure Interactions.* Models, Analysis and Finite Elements.

119. M.L. Bittencourt, N.A. Dumont, J.S. Hesthaven (eds.), *Spectral and High Order Methods for Partial Differential Equations ICOSAHOM 2016.* Selected Papers from the ICOSAHOM Conference, June 27-July 1, 2016, Rio de Janeiro, Brazil.

120. Z. Huang, M. Stynes, Z. Zhang (eds.), *Boundary and Interior Layers, Computational and Asymptotic Methods BAIL 2016.*

121. S.P.A. Bordas, E.N. Burman, M.G. Larson, M.A. Olshanskii (eds.), *Geometrically Unfitted Finite Element Methods and Applications.* Proceedings of the UCL Workshop 2016.

122. A. Gerisch, R. Penta, J. Lang (eds.), *Multiscale Models in Mechano and Tumor Biology*. Modeling, Homogenization, and Applications.

123. J. Garcke, D. Pflüger, C.G. Webster, G. Zhang (eds.), *Sparse Grids and Applications - Miami 2016.*

124. M. Schäfer, M. Behr, M. Mehl, B. Wohlmuth (eds.), *Recent Advances in Computational Engineering*. Proceedings of the 4th International Conference on Computational Engineering (ICCE 2017) in Darmstadt.

125. P.E. Bjørstad, S.C. Brenner, L. Halpern, R. Kornhuber, H.H. Kim, T. Rahman, O.B. Widlund (eds.), *Domain Decomposition Methods in Science and Engineering XXIV*. 24th International Conference on Domain Decomposition Methods, Svalbard, Norway, February 6–10, 2017.

126. F.A. Radu, K. Kumar, I. Berre, J.M. Nordbotten, I.S. Pop (eds.), *Numerical Mathematics and Advanced Applications – ENUMATH 2017.*

127. X. Roca, A. Loseille (eds.), 27th International Meshing Roundtable.

128. Th. Apel, U. Langer, A. Meyer, O. Steinbach (eds.), Advanced Finite Element Methods with Applications. Selected Papers from the 30th Chemnitz Finite Element Symposium 2017.

129. M. Griebel, M. A. Schweitzer (eds.), Meshfree Methods for Partial Differential Equations IX.

For further information on these books please have a look at our mathematics catalogue at the following URL: `www.springer.com/series/3527`

Monographs in Computational Science and Engineering

1. J. Sundnes, G.T. Lines, X. Cai, B.F. Nielsen, K.-A. Mardal, A. Tveito, *Computing the Electrical Activity in the Heart.*

For further information on this book, please have a look at our mathematics catalogue at the following URL: `www.springer.com/series/7417`

Texts in Computational Science and Engineering

1. H. P. Langtangen, *Computational Partial Differential Equations.* Numerical Methods and Diffpack Programming. 2nd Edition
2. A. Quarteroni, F. Saleri, P. Gervasio, *Scientific Computing with MATLAB and Octave.* 4th Edition
3. H. P. Langtangen, *Python Scripting for Computational Science.* 3rd Edition
4. H. Gardner, G. Manduchi, *Design Patterns for e-Science.*
5. M. Griebel, S. Knapek, G. Zumbusch, *Numerical Simulation in Molecular Dynamics.*
6. H. P. Langtangen, *A Primer on Scientific Programming with Python.* 5th Edition
7. A. Tveito, H. P. Langtangen, B. F. Nielsen, X. Cai, *Elements of Scientific Computing.*
8. B. Gustafsson, *Fundamentals of Scientific Computing.*
9. M. Bader, *Space-Filling Curves.*
10. M. Larson, F. Bengzon, *The Finite Element Method: Theory, Implementation and Applications.*
11. W. Gander, M. Gander, F. Kwok, *Scientific Computing: An Introduction using Maple and MATLAB.*
12. P. Deuflhard, S. Röblitz, *A Guide to Numerical Modelling in Systems Biology.*
13. M. H. Holmes, *Introduction to Scientific Computing and Data Analysis.*
14. S. Linge, H. P. Langtangen, *Programming for Computations* - A Gentle Introduction to Numerical Simulations with MATLAB/Octave.
15. S. Linge, H. P. Langtangen, *Programming for Computations* - A Gentle Introduction to Numerical Simulations with Python.
16. H.P. Langtangen, S. Linge, *Finite Difference Computing with PDEs - A* Modern Software Approach.
17. B. Gustafsson, *Scientific Computing from a Historical Perspective.*
18. J. A. Trangenstein, *Scientific Computing.* Volume I - Linear and Nonlinear Equations.

19. J. A. Trangenstein, *Scientific Computing*. Volume II - Eigenvalues and Optimization.

20. J. A. Trangenstein, *Scientific Computing*. Volume III - Approximation and Integration.

For further information on these books please have a look at our mathematics catalogue at the following URL: `www.springer.com/series/5151`

Printed in the United States
By Bookmasters